P9-DNI-082

SAVING AMERICA'S WILDLIFE

24.95

THOMAS R. DUNLAP

Saving America's Wildlife

PRINCETON UNIVERSITY PRESS

333. 95
D

Copyright © 1988 by Princeton University Press

Published by Princeton University Press, 41 William Street,
Princeton, New Jersey 08540
In the United Kingdom: Princeton University Press, Guildford, Surrey

All Rights Reserved

Library of Congress Cataloging in Publication Data will be found
on the last printed page of this book

ISBN 0-691-04750-2

This book has been composed in Linotron Baskerville

Clothbound editions of Princeton University Press books are printed
on acid-free paper, and binding materials are chosen for strength
and durability. Paperbacks, although satisfactory for personal
collections, are not usually suitable for library rebinding

Printed in the United States of America by Princeton
University Press, Princeton, New Jersey

For

Susan Laura Miller

my wife
for everything

Contents

$$\frac{3}{2} + \frac{5}{2} =$$

Preface

This book is about American nature myths in the last century. The word "myth," I hasten to say, is not used here in a pejorative sense. It happens to be the best term for those descriptions of the world every society provides to the people growing up in it, descriptions that locate them in the world and within their society. Henry Nash Smith, in *Virgin Land*, described a myth as "an intellectual construction that fuses concept and emotion into an image."[1] That will do: myths are shorthand, the things we never learned but we all know. Every culture has its own set, and conspicuous among them are the ones explaining the natural world and humans' relation to it. In that respect we are just like the ancient hunters who painted pictures on cave walls or the Middle Eastern farmers who first erected temples to their gods. We differ from them in the ways we transmit our myths, the authority we invoke to justify them, and the speed with which we change them. They told stories around the campfire; we have books, magazines, movies, and television. They appealed to the priest of the local god; we seek out scientists. During their lifetime the world and ideas changed little; ours shifts so rapidly that one generation hardly understands another.

Compare our current ideas about nature and wild animals with those widely accepted in the 1850s. Imagine Americans of that period listening to our debates about wildlife and wilderness. What would they think of the battle over the snail darter and the Tellico dam or the continuing controversy over irrigation on the North Platte and its effects on the migrating sandhill and whooping cranes? Coming from a society that was seeking to kill every wolf on the continent, what would they make of our plans to reintroduce the timber wolf into parts of its old range? These policies and the reasons behind them would be as foreign as the philosophy of the ancient Chinese.

They lived in a world in which the almost unchallenged view was that the world had been made in six 24-hour days by a god who had made humans the crown of creation and given them dominion over nature. People thought of animals as "dumb brutes," without thoughts, emotions, or the capacity to feel pain. Wildlife was "good" or "bad," "useful" or "worthless" as it suited human needs and fit a human order. It was there to be used, and to be used up. Today we see ourselves as the product of an apparently random process that took billions of years. We believe that animals have feelings and can in some way think. A significant minority among us grants animals rights and seeks to have society do the same. We "instinctively" draw back from cruelty and pain. We see species as interdependent parts of the ecosystem. Each, we believe, has value in sustaining the whole.

What brought us from poisoning "varmints" to reintroducing wolves, from seeing the world as a set of resources to regarding it as a "web of life?" Industrialization certainly played a part. It made us "safe" from natural disasters. It put wild nature at our mercy. But it is not decisive. Nor is the heritage of Western civilization or Americans' experience of "conquering the wilderness."[2] These provided the materials for our modern nature myths. It is science that made the plans and directed the construction. In the nineteenth century evolutionary biology told us of our origins and showed us how we were connected to other forms of life. It challenged, then displaced, religion as the source of knowledge about the world. In the twentieth century, ecology provided a complementary vision. It explained how organisms interacted, how the "economy of nature" really worked. Science has been the guide—at least rhetorically—for wildlife and nature policies and for the picture of nature we receive in books, movies, and television shows. It justifies Romantic identification with nature and animal rights. It is, in short, the authority to which we turn for new ideas and ratification of old ones.

Whether our faith in science is well placed is another question. Do scientists really understand as much as we think they do? How well do we understand the ideas to which we appeal? No matter. As a society we have loaded on this particular social activity the burden of knowledge. We have made its practitioners our oracles. How did this happen, and what have the results been?

Our story begins in the last half of the nineteenth century, with the end of the frontier, the rise of Darwinism, and the professionalization of biology. The rapid disappearance of unsettled land marked the end of an era, and if the frontier had shaped American virtue, what would the country do without it? Americans had to decide what place the remaining wilderness and its wildlife would have in their new industrial civilization. While they grappled with nature's place, they also had to decide what place they had in nature, for Darwinian evolution was destroying accepted and comfortable ideas about humans' status. Less obvious but equally important, biology was becoming an academic science. People were building an institutional structure that would support research and forming a self-conscious community devoted to the study of the field. Science would now affect us less by the individual ideas of genius than by the theories that came from this community— which would itself shape the next generation of workers.

Between the two world wars ecology built quietly on these foundations. Animal ecology, which had been a poor relation to plant studies, found concepts suited to its research problems—ecosystems, food chains, trophic levels, and niches. It began to show us just how the natural world was organized. Game managers, busily engaged in turning their field into a rigorous applied science, used ecological ideas in their research. Nature writers spoke less of individual animals, more of the "web of life" and its wonders. By the time World War II diverted attention and people from wildlife concerns, ecology was well established and beginning to affect at least the nature enthusiasts among the population.

After World War II the ideas of ecology (as various people have conceived them) reached the public in books, articles, movies, television, and formal education. Conscious of our effects on the environment and the dangers that human action posed to it and to us, we Americans embarked on a crusade to clean up the land and to save nature. In the late 1960s "ecology" became a battle cry for reform, and in a burst of environmental enthusiasm (approximately 1966 to 1973) Americans turned these ideas into law. We are now dealing with the consequences of those decisions, making the choices that will reconcile our desire for a complete ecosystem with our other needs.

Here I tell this story through a study of wildlife policy and attitudes toward wild animals, with emphasis on wolves and coyotes. This may seem a roundabout path, but it is a direct one. Of all the things we find in nature, animals have fascinated us the most, and they have been important elements in our cultures since the days of stone and bone tools. We have worshiped animals and feared them. We have traced our descent from them and defined ourselves by the gulf between them and us. We have made them symbols of vice and virtue. Animals that lived by killing others have had a special place in many cultures. They were symbols of courage, strength, ruthlessness, and destruction.

Western civilization has been no exception. It has attached enormous emotional and symbolic weight to predators, particularly to wolves. The animal was for centuries the symbol of the feared and hated wilderness and, in the legend of the werewolf, it came to stand for the evil in humans. The settlers gave the wolf in the New World the same attributes it had had in the Old, and their descendants attached to the coyote the European fox's legendary craft and cunning. When Americans set about saving nature, wildlife, and wild lands in the late nineteenth century, they made exceptions for the "bad" animals—the predators—and chief among them were the wolf and the coyote.[3]

Today, the Endangered Species Act has replaced the bounty laws that crowded state statute books. Instead of the "despicable," cowardly wolf, ready to turn on its wounded packmates or devour its own dead, we are told of wolves living in a natural world where all species have a function, where the death of prey animals is "the predators' gift to the land." Now it is man who is "the killer, the slayer, the luster-for-blood—[who] has sought to expurgate himself of his sin and guilt by condemning the predatory animals."[4] The coyote has been rehabilitated too. The sly, sneaking "varmint" is now "God's Dog," a valued part of the ecosystem. Poisoning is outlawed; official policy is to "remove the offending individual animal"; and researchers work on ways to "reduce coyote-sheep interaction."[5]

We are now reintroducing the wolf and its smaller relative, the red wolf, into areas where, fifty years ago, we sought to exterminate them. (The coyote, on the other hand, has been holding its own and needs no such aid.) In Wisconsin the state's Department of Natural Resources is intensively studying the few wolves that have crossed the border from Minnesota. Biologists are watching, and watching over, wolves that in 1985 crossed from Canada into Glacier National Park. In North Carolina the U.S. Fish and Wildlife Service is preparing to release into the southern wilderness eight red wolves bred in captivity and the pups they produced—the first contingent of what the Service hopes will be viable populations of the red wolf in the South.

AT THIS POINT non-historians might as well leave by the door on the right, the one marked "To Chapter One." The rest of this preface is intended for friends and colleagues. Some general words for my fellow historians first. This book, in part deliberately, does not follow most of the well-worn paths in our field. It is recognizably environmental history, but it pulls together elements not normally found side by side—nature literature and bureaucratic politics or ecology and the humane movement, to name two odd combinations. My friends in policy studies find this untidy. Historians of ecology have been equally appalled. People in several disciplines will loudly object that I have told only part of the story. True, but no book can tell the whole story of such a complicated change. This one uses the revolution in Americans' ideas about wolves and coyotes in the last fifty years to illustrate the larger themes of science and ideas, and the changes in ideas, about nature. The venture into wildlife policy, Darwinian evolution, ecology, hunting, nature appreciation, and nature stories are meant only to answer the question: How did we go from hating these animals to loving them, and what does this about-face tell us about our society?

The perceptive will have other complaints. My argument that there is a progression from ideas to organizations that implement them and finally to public policy will be as unwelcome to some as the way I have cut across other people's academic pastures; it looks too much like intellectual history (an unfashionable area these days). Besides some observations on Aldo Leopold, whose career and ideas I use to illustrate and exemplify a shift in ideas in the 1930s, there is little here on the Great Men of Environmental History, and my discussion of policy rarely rises to what we normally consider the important levels. Finally, I only skim the surface of science and introduce but never fully develop scientific concepts.

All this reflects my judgment of what is important. Historians have, I believe, too often neglected the day-to-day activities that make policy. We have looked instead at the process by which laws are made, and we have seen legislative action as decisive. It is not; laws are often what the bureaucracy makes them. We have too often characterized "government" as an inert lump that must be moved by (usually high-minded and pure-souled) conservation organizations, and we neglect the complex interplay of people within and without the government and the ways in which agencies have roused public support—which then "forced" them to take action. There is an anti-institutional bias in much historical work, somewhat offset by the attention we lavish on a few agencies and private organizations. The same is true of individu-

xiii

als. Great men and women are indeed great, but many others support them, and often they rely on institutions and are shaped by them.[6] As for science: what scientists thought and how they evolved and changed their ideas is less important (for this study) than the broader picture that came to the public.

How much of this structure I brought to my work and how much was formed in the course of research I cannot say. What became clear to me was that we have barely begun to understand how our ideas about nature have evolved and changed as we have moved from a rural, agricultural society to an urban, industrial one, and as we have developed a much more quantitative and detailed picture of how the natural world around us works. In doing the research for this book, I have been struck by the lack of information on many of the questions that occurred to me. Admittedly it was also gratifying to find some gaps, and I have tried to fill as many of them as I could. I hope I have succeeded.

Blacksburg, Virginia
September 1987

Acknowledgments

The obligations incurred in writing a book are one of the hardest debts to repay. No one keeps accounts, and a passing comment can be as valuable as extensive criticism. I have tried to steer a middle course between the impossible task of thanking everyone and the discourteous one of neglecting those who have made substantial contributions. To those I have wrongly omitted, I offer my apologies.

For financial assistance I thank the History Department, the College of Arts and Sciences, and the Center for Programs in the Humanities of Virginia Polytechnic Institute and State University. A summer seminar in western history funded by the National Endowment for the Humanities gave me a rewarding summer in Davis, California, with W. Turrentine Jackson and a varied and interesting group of people. A grant (SES-8319362) from the National Science Foundation and a sabbatical from Virginia Tech allowed me a year to write, free of teaching and committee work.

I did much of my work at the National Archives and Records Service. I thank the staff, especially Renee Jaussaud. Her enthusiasm and hard work have made it easier for many of us to work in wildlife and natural resources records. Others who helped include the staff members of the Bancroft Library at the University of California, Berkeley; the library at the University of California, Davis; the Manuscript Division, Library of Congress; the Conservation Center of the Denver Public Library; the University of Wisconsin Archives in Madison; the American Museum of Natural History in New York; and the Iowa State University library system. I appreciate the permission of the Bancroft Library to quote from an oral interview in their records. I thank *Pacific Historical Review* for permission to use in Chapters Four and Eight material that appeared in that journal as "Values for Varmints: Predator Control and Environmental Ideas, 1920–1939," *Pacific Historical Review*, 53 (May 1984), 141–161, and "American Wildlife Policy and Environmental Ideology: Poisoning, Coyotes, 1939–1972," *Pacific Historical Review*, 55 (August 1986), 345–369.

The students in my American Environmental History class listened to various versions of my ideas with interest and the generous patience they extend to teachers who show real interest in their subject. In almost all of my professional talks and articles in the last six years I have tried out some of the ideas that appear here. I thank the commentators, referees, and editors. A special mention is due to Al Runte, who has given generously of his time to criticize my work. The late A. Starker Leopold discussed his father's ideas with me and gave me access to his archives and permission to use his father's papers. David Wake, director of the Museum of Vertebrate Zoology in Berkeley, permitted me to use the records there. The late Dr. Ruth Risdon Storer gave me access to her husband's field notes. E. Raymond Hall chatted with me and gave me papers and access to his files. Ronald Nowak of the Office of Endangered Species has helped with papers, information, and published work. Robert Rudd gave me some useful leads and introduced me to one of his students, Angus McIntyre, who made other contributions. John Gottschalk, Jack Berryman, John Grandy, and Thomas Kimball spoke to me about their parts in this story. My thanks to all these people.

My fellow historians have been most helpful. At Virginia Tech I owe special debts to Harold Livesay, Larry Shumsky, and the people at the departmental seminar. I have used Frank Egerton's scholarship and bibliographic work and benefited from several years of infrequent but enlightening letters; I much appreciate his generosity. Keir Sterling provided sources, comments, and information. Donald Worster suggested I look at some of the records in Denver, and I have leaned on and disputed some of his work. Morgan Sherwood's work and comments were helpful. Lisa Mighetto told me a great deal about the humane movement and its relation to wildlife policy. My thanks to them.

I have dedicated this book to my wife, Susan Laura Miller, because of all that she has done (and a few things she has refrained from doing) through our years of married life. Our daughter, Margaret Miller Dunlap, has helped in her own way. Her birth pressed me to get the manuscript of my previous book off to the publisher. While I worked on this one, she provided distraction, relief, and a sense of nonscholarly realities. She fully deserves the academic child's privilege of seeing her name in print.

SAVING AMERICA'S WILDLIFE

PART I

Foundations for a Wildlife Policy

In the last part of the nineteenth century and the first two decades of the twentieth, Americans laid the groundwork for a wildlife policy that would be more than a set of laws regulating the taking of food from the woods. They found new virtues in wilderness and wildlife. They set about saving some places and some wildlife. In response to the chaos of the Darwinian struggle, they evolved a new myth that reconciled science and Romanticism. Less obvious but no less important, they put in place institutional and bureaucratic foundations for wildlife policy. They founded academic departments, museums, and government agencies. They developed organizations that could study nature and affect policy toward it.

Paradox and irony abound. Magazines printed on high-speed presses, distributed by a national transportation network, and supported by advertising preached sport hunting and nature appreciation, taught the secrets of woodcraft, and extolled the virtues of the frontier. Government agencies, claiming the authority of science, were concerned to save the wildlife that was the heritage of the "primitive" frontier. One of the most sophisticated intellectual creations of Western civilization—modern science—told people how to view and react to that most basic of experiences, the biological world around us. None of this created an effective science of wildlife management or had much effect on basic attitudes toward wild animals. Only with time would these developments be seen as the beginning of a new period.

CHAPTER ONE

Saving Animals for Use

1880-1910

The good Lord put us here and the Good Book says, "man shall have dominion over all creatures." They're ourn to use. [1]

The man who said this expressed not merely the "sentiments of the great majority of the people he knew," but the sentiments of most Americans. The white people who settled North America saw wild animals as objects to be used and used up. Some were food, others furnished fur, and these the settlers hunted or bought from the Indians. Others they killed or drove off the land as part of the process of turning the wilderness into farms and pastures. There were no thoughts of sportsmanship or conservation in these acts. Wild animals, even the useful ones, were temporary resources. They would, like the Indian and the forest, vanish before advancing "civilization."

Some creatures the settlers sought to exterminate. Chief among them were predators, such as the wolf, that preyed on livestock and were thought to kill people. The smaller carnivores—weasels and foxes—might be tolerated or killed when chance offered. Against the wolf and its like it was war to the knife. Government took a hand; Massachusetts Bay and Virginia began paying bounties for wolf scalps in the 1630s, and the practice remained an almost universal feature of state statute books into the twentieth century. [2] Even without this incentive people killed wolves. The early colonists used pits and dead-falls. They organized "circle drives," herding game and predators to a central location to be killed. They put out set guns and poisoned meat, buried fishhooks in balls of fat, and made leg and head snares. In the early nineteenth century their descendants added strychnine and steel leg-hold traps. [3] By the 1850s wolves were rare in the East, and the red wolf (which the settlers did not distinguish from its larger relative) had probably been wiped out over much of its range.

The slaughter increased after the Civil War. Repeating rifles, steel traps, strychnine, and a national market served by railroads made it possible, and profitable, to kill every animal that could be found. Professional "wolfers," carrying bottles of strychnine sulfate, followed the buffalo hunters, using the carcasses left behind as bait. After them came the ranchers, with more poison. Two wildlife biologists, both active in federal predator control for years, summarized the results:

Destruction by this strychnine poisoning campaign that covered an empire hardly has been exceeded in North America, unless by the slaughter of the passenger pigeon, the buffalo, and the antelope. There was a sort of unwritten law of the range that no cowman would pass by a carcass of any kind without inserting in it a goodly dose of strychnine sulphate, in the hope of eventually killing one more wolf.[4]

America was to become the "Garden of the World" and there was no place in this pastoral Eden for the wolf. Its death was, in fact, a sign of civilization and progress.[5]

Roots of Change

The large animals might all have vanished, and many of the smaller ones with them. The rural landscape might have been given entirely over to fields, farms, woodlots, and pastures. That did not happen. Even as the slaughter reached its peak, a significant number of Americans ceased to see wildlife as an obstacle or a temporary source of food and money. There had always been a few, beginning with the naturalists of colonial America, but in the 1880s there was, for the first time, a movement for wildlife preservation. It did not seek to save all animals, and it did not espouse the ideas now popular among wildlife's defenders, but it was a beginning.

There were two parts to this enthusiasm. One was sport hunting, which found in the chase an arena for forming and testing the character of Americans that would substitute for the now vanishing frontier. Later generations, going to the field, could re-create the pioneer experience and develop the virtues of the pioneers. To accomplish this, wildlife, or at least the species that were quarry for the chase, had to be saved. The other part was nature appreciation, an offshoot of Romanticism. Wild animals, nature lovers believed, provided an opportunity for spiritual and aesthetic experiences. Contact with them, like appreciation of beautiful scenery, was an antidote to the artificial life of civilization. For both groups, wild animals now appeared as part of American history and culture, a precious legacy we had to preserve.[6] Neither hunters nor nature lovers were yet concerned to save the wolf or the coyote (just then becoming a target of poison and traps), but their movements contained the seeds of change.

It was the destruction of the buffalo herds that made wildlife protection a public issue. The endless herds had been a feature of Plains travel and comment since Coronado had found them grazing on the Kansas prairies in 1541. The herds had seemed endless and endlessly productive, but in the 1870s breech-loading rifles, the Army's enthusiasm for starving the Plains' tribes into submission, and the railroad spelled their doom. For a few years the grasslands resounded with gunfire, as if to the sound of battle. Then there was silence.

The first reaction was more reflexive than reflective. Many people, including the young Theodore Roosevelt, rushed off to get one of the few remaining "trophies."[7] The Smithsonian Institution checked its inventory and sent out an expedition to get more specimens for its collections. Judgment came later. Some, including the conservationist editor of *Forest and Stream*, George Bird Grinnell, lamented the loss but thought the buffalo's passing "a necessary part of the development of the country."[8] Others were less forgiving. William Hornaday, taxidermist, head of the New York Zoo, ex-hunter, and passionate wildlife defender, blamed the "descent of civilization, with all its elements of destructiveness, upon the whole of the country inhabited by that animal . . . and . . . the reckless greed . . . wanton destructiveness [and] . . . improvidence in man's husbanding such resources as come to him from the hand of nature already made."[9]

People rallied to save this symbol of the West. A few ranchers had rounded up stray survivors and begun private herds. There were a few buffalo in Yellowstone National Park and in zoos around the country. Grinnell, despite his feeling that the buffalo had to pass from the scene, began a campaign in *Forest and Stream* to save the buffalo. He fought to end hunting in Yellowstone Park and to keep the species as part of the West. In 1905 Hornaday and his friends organized the Bison Society. They made an inventory of the remaining animals, provided aid and information for private owners, and lobbied Congress for a refuge. When Congress did set up a national refuge, Hornaday's New York Zoo stocked it.

No one rallied to save the buffalo wolf. Still, there was in this increased concern for wildlife and wild nature the beginning of a place for all wildlife, including the despised "vermin." The people wanted

7

to retain and encourage the old skills of woodcraft, and movements as diverse as women's clubs, sportsmen's associations, the Boy Scouts, and charitable societies that sent slum children to summer camp saw in nature a refuge from civilization and a test of character. As people invested the activities of the pioneers, even the simple act of being in "wild" country, with new qualities, they had to face, at least implicitly, the question: Could the wilderness and wilderness experience be saved without the wildlife of the wilderness, without the wolf?

More than nostalgia lay behind the new attraction to wilderness and the appeals to frontier virtue. Sport hunting, nature study, and the "strenuous life" were most popular among the descendants of the pioneers—the white, "Anglo-Saxon," Protestant people who dominated American society—and their commitment was in part an effort to preserve their culture and its virtues in a new world. The next generation would grow up in cities, not on farms, and it would consist in large part of immigrants and immigrants' children. It would have no contact with the old civilization. If the virtues formed in the struggle with the wilderness—the virtues that had made America what it was—were to continue to be the bedrock of American life, the new generation would have to be educated to appreciate nature and "manly sport with a rifle."

Hunting as a Sport

In hunting the finding and killing of the game is after all but a part of the whole. The free, self-reliant, adventurous life, with its rugged and stalwart democracy; the wild surroundings, the grand beauty of the scenery, the chance to study the ways and habits of the woodland creatures—all these unite to give the career of the wilderness hunter its peculiar charm. The chase is among the best of all national pastimes; it cultivates that vigorous manliness for the lack of which in a nation, as in an individual, the possession of no other qualities can possibly atone.[10]

The transformation of hunting was one of the first steps toward wildlife preservation. Hunting had been a way of getting meat for the table; people used whatever means came to hand and killed as much as they could. Under the influence of English ideas about field sports and concern to preserve pioneer virtues, hunting became a ritual activity. Its aim was now not meat or fur but a "fair chase" and a good, "manly" character. The hunter killed only "game" species by "fair"

8

methods. He acted in the field according to the "sportsman's code." This made animals, at least game animals, a resource. They would be killed, but the stock had to be preserved and renewed. It is too much to claim (as one historian has done) that hunting was the cradle of the conservation movement but hunters were the first organized group to press for wildlife preservation and until well into the twentieth century provided the money for wildlife work.[11]

The idea of hunting as a sport came to America from England in the early nineteenth century and became popular among the urban upper class around the time of the Civil War. Its great American apostle was Henry William Herbert, an English immigrant who began writing on field sports in the 1830s under the pseudonym Frank Forester.[12] Hunting and fishing, he preached, were not ways of wasting time or getting food for the table: they were the recreations of gentlemen, who pitted their craft against that of the game in "fair chase." They were preparation for life. Herbert dwelt on the military superiority of the English upper classes, which he laid to their devotion to outdoor sports. He asked to "be permitted to doubt whether the Schottishing flower of young York" would do as well when put to the test of battle as the English had done in the Crimea.[13]

The primary benefit of field sports was their influence on character; they cultivated and tested virtue. The gentleman was guided by the ideal of sportsmanship, which lay in the "true spirit, the style, the dash, the handsome way of doing what is to be done, and above all, in the unalterable *love of fair play*, that first thought of the genuine sportsman."[14] He killed none but game animals and those only by "sporting" methods. Not for him the sitting duck, the swimming deer, or the collection of migrating songbirds. He shared shots and the bag with his partner. On no account did he mistreat his dogs or horses. Cruelty to animals, Herbert said severely, showed that a man was not a true sportsman and a gentleman.

The sportsman's care to take only "suitable" species was in sharp contrast to the usual American practice. Our ancestors ate almost anything that walked, flew, or swam. Thomas Nuttall included in his *Ornithology* (1832) notes on the preparation of species few of us would think of as food—flickers, meadowlarks, bobolinks, and cedar waxwings. A guide to the New York market (1867) gave recipes for preparing juncos. It spoke of mixed strings of meadowlarks, catbirds, and purple finches, pies of cedar waxwings and goldfinches. Songbirds re-

9

mained part of the diet, particularly in the South, until the twentieth century. Nor was this a matter of poor people eating for survival; even genteel nature lovers gave robin soup to invalids.[15] Among mammals the choice was almost as wide. Deer, bear, moose, and elk shared the bill of fare with squirrel pie, baked possum, and groundhog stew. Certain species were "higher class," but all went to someone's pot.

The sportsman's insistence on "fair chase" and on leaving some game for the next man and the next year also ran against common practice. People shot swimming deer from canoes, sent dogs after them in deep snow, and used fires or lanterns for night shooting (a primitive form of jacklighting). They fired "punt guns"—small cannons holding a half pound or more of shot, cut nails, and scrap—into rafts of "sitting ducks" and turned shotguns on passing or roosting flocks. There were hunting regulations, but bags were generous, seasons long, and there were no wardens. When Herbert began preaching sportsmanship, large game was already vanishing from the East.

By the 1850s Herbert could congratulate himself that at least some had seen the light. The respectable parts of the community had once regarded field sports as the pursuits of loafers and the lower classes. They were now coming to see, Herbert said, that all a man's attention should not go to business. They appreciated a day afield as a pleasant and invigorating relief from the anxieties of business life. There were sportsmen's clubs, at least in a few cities, and the ideal of sportsmanship was spreading. It began to spread more rapidly after the Civil War. In the 1870s sportsmen's clubs tripled in numbers to over three hundred. Most were small and local, but a few had greater influence. The most prominent was the Boone and Crockett Club, formed in 1887 by Theodore Roosevelt and George Bird Grinnell. A group of a hundred sportsmen who had taken in "fair chase" with a rifle at least three species of North American big game, the club drew on the political and social elite of America. Particularly after Roosevelt became president, the club's efforts to set up game refuges and encourage game conservation bore fruit.[16]

Forester's disciples found a new medium for their message. Inexpensive magazines—the product of high-speed printing, low-cost paper, advertising revenue, and nationwide mail delivery—brought sportsmanship to the masses. *The American Sportsman* began publication in October 1871, *Forest and Stream* in 1873, *Field and Stream* (not the modern one, but a predecessor) the next year, and *American Angler* in 1881.[17] To anyone with a few spare nickels they offered weekly or monthly doses of advice, information, and exciting stories of nature and sportsmen afield.

The sportsman's creed soon took on an American tinge. The magazines' ideal hunter was not the upper-class Englishman or American "shooting" on his own ground. He was the wilderness hunter. *Forest and Stream* declared its objects to be the promotion of a "healthful interest in outdoor recreation and . . . a refined taste for natural objects." It wished to make its readers "familiar with the living intelligences that people the woods and the fountains." It would "teach you those secrets which necessity compelled the savages to learn . . . first principles [out of which] our civilization grew but [of which] we are ignorant."[18] The American hunter, even (or perhaps especially) the upper-class hunter, was more than a good shot, more even than a sportsman. He was a woodsman, the heir of Daniel Boone and Davy Crockett.

Sporting journals also organized hunters for political action. They had to. Americans had, with few exceptions, no landed estates and little chance to take trips to Africa or India. Even if they had, shooting birds driven to the gun or big game on a safari would not provide contact with nature or re-create the pioneer experience. American hunting depended on public land and access to private land, and to a continuing supply of game on it. But the supply of game was declining rapidly. To save their sport, hunters had to create a mass movement and call in the government to save the animals. Their program was simple. "The history of American management," Aldo Leopold remarked in 1933, has been "until recently almost wholly a history of hunting controls."[19] Sportsmen wanted to outlaw "unsporting" methods that gave the game no chance: jacklighting, hunting deer with dogs or in the water, using punt guns, and baiting areas to attract game. They wanted to reduce annual kills through lower bag limits, shorter seasons, and regulations on the kind of firearms hunters could use. They sought the abolition of spring hunting. They wanted all these new laws enforced, preferably by a professional set of wardens under the direction of a state game commission.[20]

Here as elsewhere the magazines led the way. As early as 1874 Charles Halleck of *Forest and Stream* protested the slaughter of the buffalo and tried to get Congress to protect the species. His successor, George Bird Grinnell, led the fight to ban hunting in Yellowstone National Park, spring shooting, and, after 1894, market hunting. Most of this preaching was as gentlemanly as the conduct it advised, but a few people scorned the niceties. G. O. Shields, editor of *Recreation*, waged a fierce war against the "game hog." From 1896 until 1905, when he lost editorial control of the magazine, Shields printed pictures (culled from small-town papers) of outrageous bags. He excoriated the people involved and handed out numbers and places in the "hog pen."[21]

Sportsmen were most concerned with game animals, but they helped protect other species as well. With nature lovers and ladies' clubs, they opposed the killing of songbirds—though their objection was to the lack of "sport" in the pursuit. They joined the campaign against the slaughter of the plume birds in the 1890s. They played a large part in the movement for the Lacey Act, the first general federal wildlife statute. They even helped nature lovers organize. In 1886 George Bird Grinnell proposed in *Forest and Stream* the formation of a club, to be called the Audubon Society, for those interested in protecting and enjoying birds. Public response was so great he had to abandon the club three years later as too great a drain on his time. Others, though, picked up where he left off.[22]

By the beginning of the twentieth century the states, which had primary responsibility for wildlife, were acting on the sportsmen's program. They were setting lower bag limits, reducing the length of open season, passing license laws, establishing game commissions, and hiring wardens.[23] By 1920 regulation of some sort would be almost universal. The hunters' success owed much to public interest, but it was also due to the social composition of the movement. Its leaders and many of the followers were men of the governing class and dominant ethnic group—white, Protestant "Anglo-Saxons." Hunting, in fact, often drew class and racial lines. William Hornaday was not alone in inveighing against "negro and Italian bird-killers." He disclaimed any racial prejudice (he was sure that he would be accused of it), but "[n]o white man calling himself a sportsman ever indulges in such low pastimes as the killing of [songbirds] for food." Shame, he cried, on such people and on the legislatures that allowed it![24] Some, apparently, felt the shame. States outlawed birdlime and nets (the favored methods of Italian immigrants) and barred aliens from carrying or owning firearms.[25]

In the long run the sportsmen's work for wildlife would help even the despised wolf and coyote. The mystique of the wilderness could hardly be sustained without the animals that symbolized the wild, and programs set up to aid in game preservation could be extended to include other species. In the short run, though, sport hunting provided another argument for predator extermination. "Varmints" competed with the hunter for scarce trophies, did it all year round, and did not use sportsmanlike methods. The bear might be saved, for it was a game species, but not the wolf or coyote. They had to go.

Hunting was a test of male character, the field a place for male camaraderie. Women might participate in its easier activities—fishing or shooting upland birds (with, of course, a lighter "ladies'" gun)—but most found a place more "appropriate" to their "sphere" and their sensibilities. They "appreciated" nature or taught children the "proper" attitudes toward our "friends in fur and feathers." But the movement for nature appreciation had many parallels with sport hunting. Both grew out of European ideas modified for American conditions. Both appealed most strongly to the dominant social group, and both groups sought legislation to support their goals. Like hunters, nature lovers saw their cause as a way to rescue the virtues and culture of the country's rural past. The movements overlapped to some degree. They had some common goals—protection of songbirds, saving the plume birds, and setting up wildlife refuges. Hunters extolled nature appreciation as one of the benefits of hunting, and some hunters, like Theodore Roosevelt, were avid bird watchers and enthusiastic nature lovers.[26]

The primary vehicle for nature enthusiasm was the nature essay or story. From the 1870s these began appearing with increasing frequency in general magazines.[27] The pieces ran the gamut from sentiment to popular science, the subjects from backyard birds to wilderness wildlife. There were some common themes. Romanticism was most prominent. To study animals, all these authors said, was to know a greater reality that would lift one out of the petty concerns of daily life. It was also a democratic pursuit. One did not need to be rich or well traveled. One did not have to go to far places. Even in the cities one could see flowers, animals, and birds. Beyond this, the nature writers of the period presented a more personal view of animals. They were more than objects of appreciation: they occupied their own world, had their own lives; they were creatures like us, entitled—in a sense that the authors did not closely define—to fair treatment.[28]

In the 1890s the slaughter of birds to decorate hats and dresses mobilized nature lovers. It changed what had been a hobby and a pastime into a crusade. When fashion decreed that birds and bird feathers were stylish, birds suffered. The hardest hit were herons and egrets, whose breeding plumes were prized decorations for hats. But many other birds caught the fashion trade's attention—songbirds, hawks, and birds of paradise—and they were used in strange ways. Wings, tails, even whole, stuffed birds took their places on hats.[29]

The fight against the feather trade united scientists, nature lovers, and hunters. The U.S. Department of Agriculture's Bureau of Biological Survey and the American Ornithological Union's Committee on Bird Protection provided scientific evidence and expert testimony. Sportsmen's magazines mobilized their readers. Local Audubon societies and ladies' clubs organized other parts of the community. Humanitarian sentiment reinforced aesthetic interest. Herons and egrets were in breeding plumage during the nesting season, and hunters simply went to the rookeries and shot all the adults, leaving the young to starve.[30] A sympathetic congressman, Representative John Lacey of Iowa, sponsored a bill to end the practice, and in 1900 Congress passed it. Because the federal government had no explicit power to control wildlife the Lacey Act worked by indirection. Relying on Congressional control over interstate commerce, it banned the shipment across state lines of wild animals killed in violation of state laws. The expedient was a bit clumsy, but with the enthusiastic aid of volunteers it worked. Within a decade it largely choked off the traffic in plumes.[31]

The antifeather campaign relied on clubs and societies, but it also stimulated their growth. Through this formative period the main effort of the movement was to end the traffic in bird feathers. In 1896 William Dutcher and a group of other bird enthusiasts had revived the Audubon Club idea that Grinnell had dropped seven years before. Their group, the Massachusetts Audubon Society, was the forerunner of a national movement. In 1899 the society acquired a journal, *Bird-Lore*, edited by Frank Chapman of the American Museum of Natural History in New York. Six years later the state organizations banded together as the National Association of Audubon Societies.

Nature lovers, like hunters, wanted to make their ideals and their program part of society. But where the hunters relied on laws and magazines, nature lovers depended on education. They aimed to foster a love of nature through an understanding of wildlife, inculcated at an early age. From its beginning *Bird-Lore* had a children's section, with pictures, stories for younger readers, essay contests, and bits of natural history. The society sponsored school bird-study clubs, published plans for birdhouses, and gave schools posters and instructional materials. The society also sought to give nature study a formal place in the schools. Allied with ladies' clubs, it campaigned in state legislatures to incorporate Audubon programs into elementary-school curricula.

Like hunting, nature appreciation reflected the culture of "Anglo-Saxon" America against that of the new immigrants and the values of the middle class against those of the "lower" orders. In 1901 *Bird-Lore* rejoiced that stuffed birds and wild bird feathers were now found only on the "wearer of the molted garments of her mistress" or the " 'real loidy' who . . . with hat cocked over one eye, pink tie, scarlet waist, [and] sagging automobile coat . . . haunts the cheaper shops . . . and in summer rides a man's wheel, chews gum, and expectorates with seeming relish."[32] Schoolteachers were gently warned that "in appealing to the average child of the public school, it should be remembered of how many races this average child is compounded—races with instincts concerning what are called the lower animals, quite beyond the comprehension of the animal-loving Anglo-Saxon."[33]

The animal-loving Anglo-Saxon, however, had no more use for predatory animals and birds than did the hunter. Nature appreciation would, in the long run, work for preservation of the wolf, the coyote, and other "bad" animals. In the short run, however, genteel nature appreciation reinforced prejudice. Predators were "cruel." They "murdered" the "innocent" deer and songbirds. People in the humane movement had the most rigid attitudes. Those edging toward vegetarianism as moral principle found predation an embarrassing phenomenon.[34] Even those less extreme found predators a distasteful necessity at best. They might be needed to eliminate the suffering or unfit, but otherwise they had no place. On refuges the Audubon Society cheerfully recommended killing hawks, owls, foxes, and others that would break the peace.[35] Even scientists shared, or were willing to use, these views. Joseph Grinnell was a leader in the fight against predator poisoning in the 1920s and a strong advocate for wildlife, but he could as late as 1915 defend shooting songbirds for museum collections with the claim that the "average collector can and does on all occasions destroy Cooper and sharp-shinned hawks, and in this way certainly makes up for the small birds he shoots."[36]

Perhaps it was the specter of the Darwinian struggle for existence or uneasiness about the dark underside of industrial America, but turn-of-the-century naturalists loathed predators even more than their predecessors had. Thomas Nuttall, writing in 1832, could describe the predatory birds as being formed for their "destiny." He found the peregrine beautiful, and admired the eagle and osprey (though he did credit stories of eagles carrying off children).[37] Alexander Wilson, describing in 1808 a black eagle robbing an osprey of his catch, placed this within a human context, but in a deprecating fashion. Our sympathy, he said, "on this, as on most other occasions, generally sides with the honest and laborious sufferer, in opposition to the attacks of power, injustice, and rapacity—qualities for which our hero is so generally notorious and which, in his superior, *man*, are certainly detestable." He then added that "As for the feelings of the fish, they seem altogether out of the question."[38] Seventy years later William Hornaday said that the peregrine falcon looked best "in collections"; Cooper's and sharp-shinned hawks should be "shot on sight"; the great horned owl was an "aerial robber and murderer."[39] Even the Audubon Society, defender of all that had feathers, made an exception for the bird-eating hawks.[40]

Wolves suffered a similar fate; though never exactly popular, they lost even more ground in the late nineteenth century. John Godman in 1828 said that the wolf was "possessed of great strength and fierceness, and is what is generally called a cruel animal," but he did not condemn it outright. His contemporary, Richard Harlan, said of the genus *Canis*: "Omnivorous; voracious; very intelligent; for the most part uniting in troops in order to hunt down the more peaceable animals on which they prey. . . . the female, in a savage state, bringing forth from three to five in a birth, which they nurse with great affection and defend with courage."[41] Hornaday in 1904 described wolves as the most "despicable" of all wild animals on the continent: "There is no depth of meanness, treachery, or cruelty to which they do not cheerfully descend. They are the only animals on earth which make a regular practice of killing and devouring their wounded companions, and eating their own dead."[42] Theodore Roosevelt, who was enthusiastic about most animals, agreed. The wolf, he said, was "the archetype of ravin, the beast of waste and desolation."[43]

New and more favorable ideas about predators and predation depended on more than an enthusiasm for wilderness. They required a
picture of the world in which predators served a function, in which
predation was more than the clearing away of dying animals. That was
coming. Darwinian evolution presented a new picture of the world—
though, like hunting and nature appreciation, it was not readily apparent that it would change ideas about wolves. Biology, including the
subdiscipline of ecology, was becoming an established academic, scientific specialty. A community of scientists was forming in colleges,
museums, and state and federal agencies. Even as Americans began to
appreciate wildlife and the wilderness, their conceptions of it were
being changed and institutions were developing that would generate
newer ideas.

CHAPTER TWO

Science and the New American Nature Myth

1880-1910

There is no fundamental difference between man and the higher mammals in their mental faculties. . . . Even the lower animals . . . manifestly feel pleasure and pain, happiness and misery. . . . Only a few persons now dispute that animals possess some power of reasoning.—Charles Darwin, *The Descent of Man* (1871)

Charles Darwin shattered the world of special creation, where a Divine purpose guided everything and a gulf was fixed between man and the "beasts." He made, or seemed to make, humans into animals and the animals almost human. Nature, once the visible handiwork of a beneficent Creator, now seemed only to be the relentless struggle of all against all—devoid of order, purpose, or morality. Coming to terms with Darwin was a major task for the generation that came of age in the late nineteenth century. It was not just that the theory made nature chaotic; it was itself abstruse, incomplete, and capable of supporting a variety of interpretations. In our common descent from the animals Romantics found support for the idea that man was linked to nature—new evidence that all the world was one glorious whole. Sentimentalists and those concerned about humane treatment of animals rejoiced at evidence that we were closely related to our "dumb friends." Others pointed to the "struggle for existence" and the "survival of the fittest" to claim that we should "conquer" nature. To impose a human order on the world would prove our "fitness." Still others used the "gap" between the "civilized" and the "less evolved" races to justify colonial rule.[1]

What most people wanted was not science but sentiment, not intellectual theory but emotional identification. They were reluctant to abandon long-held ideas about order and purpose in the world. They wanted to find in nature a greater reality before which they could stand in awe. They might have abandoned science for Biblical literalism, but science was too firmly established as a source of knowledge about the world for the educated to take that way out. The solution was to reconcile Darwinism with older ideas. In this bleak vision of chaos were elements that would reinforce Victorian ideas of order, purpose, and progress. A generation searched for these with vigor.

Few people formed their ideas about nature by reading *The Origin of Species* or *The Descent of Man*. They relied on popular nature writing—explanations of Darwinism, essays on nature informed by science, and nature fiction. This had been a means of public education since the eighteenth century, but in the 1890s it took a new form: the "realistic" animal story. Here animals, presented as animals and not as disguised humans, were the central characters, observed from within. The authors of this fiction claimed the sanction of science: natural history and their own observations stood behind the incidents; animal psychology justified the mental life they attributed to their characters. The stories, though, were more than entertainment. The nature they showed, while Darwinian, was a world guided by morality and purpose. An order underlay the chaos. The stories offered a useful myth, a picture of humans and their world that incorporated knowledge and psychological needs into a satisfying whole. They made the new science serve the old Romanticism.

The new stories challenged the traditional view of animals. Because they relied on animal psychology to claim that animals thought, felt, and suffered, the stories gave weight to the movement for humane treatment. At the same time, because they faced the problems of struggle and death, they addressed, as earlier animal fiction and essays had not, the place of predators in the world. In them we see some of the first pleading for the despised "varmints." The stories show how public ideas were changing, and the debate over their accuracy, the "nature-faker" controversy, shows as well the limits of human sympathy for animals and the limits of our understanding. Out of this would come a new place for all animals, including the predators, based on a new view of animals and the world in which they lived.

The New Animal

In western civilization animals have been "dumb brutes." Tame, they were property; wild, they were *ferae naturae*, free goods until reduced to possession by capture. In either case they were things, without minds, emotions, or (many believed) the capacity to feel pain. They

19

might be made symbols of virtue or vice or, in fables and stories, be made to speak; but no one took such depictions seriously. Animals were governed by reflex and instinct, mechanical reactions to external stimuli. This may seem extreme, but consider the way in which animals have been treated in Western civilization. In medieval and early modern Europe the common amusements included bull-baiting, dog-fighting, and cockfighting. People bet on the number of rats a terrier could kill. Small children—and adults—pulled the wings off flies, drowned cats, and tormented dogs. Americans had similar amusements.[2]

Since the seventeenth century, natural history and animal taxonomy had made the point that humans were closely related to animals in body and behavior, and by the late eighteenth century this was changing society. For some, at least, the treatment of animals was becoming a moral question. Jeremy Bentham, the English philosopher, gave the issue its classic formulation in 1789. Comparing the treatment of animals and slaves, he protested that "The question is not, Can they *reason?* nor Can they *talk?* but, Can they *suffer?*"[3] Sixty years later Darwin made that a very uncomfortable question. If man's body had evolved from the "lower animals," so must his mind. If there was "no fundamental difference between man and the higher mammals in their mental faculties . . . [if even] the lower animals . . . feel pleasure and pain, happiness and misery . . . [if it was true that] animals possess some power of reason," then the question indeed became, "Can they *reason?*" "Can they *talk?*" might not be far behind.

That animals might think made animal psychology, hitherto a dead field, an exciting one. By the 1880s a school of animal psychologists was using observation and extrapolation from human mental processes to examine the mental life of animals.[4] Watching an animal, or hearing a report of what it did, the psychologist would decide what he would think under the circumstances and from this infer the animal's mental state. This was, at best, a treacherous technique (and the experimental psychologists who began work in the 1890s felt impelled to tear down the entire structure and start over), but its pitfalls, the best psychologists were certain, could be avoided.[5] The scientists' ideas of mental activity were equally crude. They conceived of two mental states: instinct, conceived of as a stereotyped reaction to a stimulus, and reason. This dichotomy, coupled with their techniques of observation, led them inexorably in one direction. When observed closely, even simple animals did things that could not be built-in mechanical reactions. But once instinct was ruled out, the conclusion was inevitable: animals could think.

This was the explicit message of popular works like W. Lauder Lindsay's *Mind in the Lower Animals* (1880). Even the lowest organisms, Lindsay declared, possessed many "advanced" mental faculties. Protozoa showed a "whole series of mental phenomena . . . will, purpose, choice, ingenuity, observation, feeling." Higher organisms were proportionally better developed. Fish were capable of "conjugal and parental love . . . [f]idelity, [s]elf sacrifice . . . [and] feelings of indignation or disgust." Birds had articulate speech and language, laws, an order of battle, judicial trials, and aesthetic taste. They could perpetuate practical jokes and appreciated domestic comfort. Dogs, or at least some of them, had a religious and moral sense. They understood language and the use of money, and had ideas of time, tune, number, and order.[6]

It is easy to dismiss Lindsay as an enthusiastic amateur—he was—but respected scientists came to very similar conclusions. George Romanes, a friend of Darwin and a careful and critical observer, believed that the higher animals shared almost all of man's characteristics. The more advanced crustaceans had reached the level of reason and had progressed in the emotional scale to the point of affection. Among the carnivores, rodents, and ruminants he found grief, cruelty, benevolence, and an understanding of mechanisms. Both apes and dogs could feel shame and remorse. They had an "indefinite morality," a state equivalent to that of a fifteen-month-old child.[7]

Even now, more than a century after *The Origin of Species*, we are fascinated by articles on dolphin squeaks and on gorillas and chimpanzees that can use sign language. Our Victorian ancestors were equally interested. Scientists wanted to find out what animals were really like; sentimentalists were anxious to find a less "brutish" nature; and pet owners were sure that Fido was "almost human." There was a flood of articles on faithful dogs, horses that could count, and elephants with remarkable memories. Magazines carried continuing debates on the extent of a dog's intelligence and whether bees and wasps were capable of gratitude.[8]

Man and the Animals

The new psychology swept the field. It also made possible the "realistic" animal story, that special form Ernest Thompson Seton and Charles G. D. Roberts developed in the early 1890s. They independently hit on the device of using animals as the central characters for nature stories, observed from within and endowed with the mental life attributed to them by animal psychology. Roberts had some trouble placing his first story, "Do Seek Their Meat from God," in 1892; editors were not quite certain how to react to this unconventional tale. But two years later scientists, naturalists, and the public enthusiastically praised the first of Seton's "scientific" stories, "The King of Currumpaw: A Wolf Story." His first collection, *Wild Animals I Have Known* (1898), went through nine printings in eighteen months. Roberts, though he never had such a spectacular single volume, sold all he could write and issued almost annual collections of his magazine stories until World War I. A host of imitators followed.[9]

Independent invention, rapid acceptance, and quick imitation suggest that the form was, so to speak, in the air. Certainly it did not come from Seton and Roberts's common background. They shared a birth year, 1860, the experience of being raised on the edge of the Canadian wilderness, and a childhood interest in nature, but little else. Roberts, the son of a New Brunswick minister, grew up in a loving family and was educated in the best schools in Canada. While still in college he was a recognized, even lionized, Canadian poet. He went from school to teaching and editing, then to a professorship. He was elected Fellow of The Royal Society of Canada in 1890. Seton, on the other hand, described his father as "the most selfish person I ever heard of or read of in history or in fiction."[10] He made his way against his father's wishes, and for years scraped out a living as an illustrator, writer of occasional pieces, and specimen collector for museums. He even worked, briefly, as a "wolfer" in New Mexico, trapping and poisoning the animals he also admired.[11]

Seton began writing nature stories, he later said, by the "archaic method, making the animals talk." Gradually he adopted the "more scientific method," in which natural history, animal psychology, and his own observations provided the material and dictated the action.[12] He had no doubts about animals' ability to think or learn. Beyond instinct, he said, was the teaching of parents and comrades, communicated by example, and the knowledge the animal got from his own experience.[13] Roberts did not go through any experiments with talking animals, but he reached the same end. The modern animal story, he said, is a "psychological romance constructed on a framework of natural history." That framework included animal psychology and the conclusions of the post-Darwinian school. We have attempted to explain animal behavior by instinct and coincidence; we have stretched these to their limits, but they have failed: "We now believe that animals can and do reason."[14]

Neither author saw science as an obstacle to sympathy with animals. On the contrary, it was a way of increasing our understanding and appreciation. "The field of animal psychology so admirably opened [by the modern nature story]," said Roberts, "is an inexhaustible world of wonder."[15]

The animal story, as we now have it, is a potent emancipator. It frees us for a little from the world of shop-worn utilities, and from the mean tenement of self of which we do well to grow weary. It helps us to return to nature, without requiring that we at the same time return to barbarism. It leads us back to the old kinship of earth, without asking us to relinquish by way of toll any part of the wisdom of the ages, any fine essential of the "large result of time."[16]

Seton agreed. In "The Kangaroo Rat," he made science the key to a new Romanticism. Finding the tracery trails of the kangaroo rats, he had been tempted to take them for fairies. "Would it not be delightful? . . . But for me, alas! it was impossible, for long ago, when my soul came to the fork in the trail marked on the left 'To Arcadie,' on the right 'To Scientia,' I took the flinty, upland right-hand path." Studying the little animals, though, he finds in them "the Little Folk, and nearer, better, and more human Little Folk than in any of the nursery books. My chosen flinty track had led me to Upper Arcadie at last."[17]

The vehicle was science; the message was moral. He wished the reader, Seton said, to learn a lesson "as old as Scripture—we and the beasts are kin . . . and [s]ince . . . animals are creatures with wants and feelings differing in degree only from our own, they surely have their rights." The stories, each showing the "real personality of the individual, and his view of life" made the point. When they did not, Seton resorted to the sermon, straight. Over the body of Redruff the partridge, caught in a snare, he orated: "Have the wild things no moral or legal rights? What right has man to inflict such long and fearful agony on a fellow-creature, simply because that creature does not

23

speak his language?"[18] Roberts, usually with more subtlety, preached from the same text. "The Aigrette" described a woman dressing for a fashionable occasion—and the slaughter at the egret rookery that provided the feathers for her hat. "The Kill" inverted the usual hunting story. Roberts told it from the point of view of the hunted, wounded, and ultimately slain moose.[19]

Both men found morals, order, and purpose in nature; indeed, this sentiment was one of the reasons these writers were popular. Seton carried this the farthest. *The Ten Commandments in the Animal World* tried to show that the laws from Mt. Sinai "are not arbitrary laws given to man, but are fundamental laws of all highly developed animals."[20] Animals that followed these laws—that were, for example, monogamous rather than polygamous—were higher in the scale of development and better fitted to survive.[21] In his nature stories the same rules applied. Little animals learned as their first lesson obedience to their mothers. Courage and endurance were as important as instinct or learning. Nature was governed by an iron law, but it was a familiar one.

Roberts was less concerned with human moral values in nature but more anxious to find order and purpose under the apparent chaos of the struggle for existence. His answer, elaborated in some two hundred animal stories, was that death was the price of life. Each lived because others died.[22] This was the theme of his first animal story, "Do Seek Their Meat from God," and the chain of death, Roberts made clear, included man as well as beast.[23] The story seems at first glance a standard melodrama. Two "panthers" (eastern mountain lions) leave their cubs in the den and set out on a night's hunting. In another part of the forest a young child has gone to play with a friend. The family has, that very day, left, and the boy, afraid to return home in the dark, now huddles in the empty cabin. The panthers hear him, and move toward the sound. The child's father, coming back from town, hears him as well. He almost passes by, thinking it is only the "squatter's brat," but moved by pity he leaves the trail and heads for the cabin. He enters the clearing just as the animals are nosing at the door. He kills them, one in hand-to-hand combat. Only then does he discover that it is not the "squatter's brat" but his own child he has saved.

What sets it apart is the point of view. As man and beasts converge on the cabin, Roberts pauses to disabuse the reader of his prejudices and drive home the moral implicit in the title.[24] "Theirs was no hideous or unnatural rage, as it is the custom to describe it. They were but seeking with the strength, the cunning, the deadly swiftness given them to that end, the food convenient to them." Nor does the story end with the father's discovery that it is his own child he has saved. It ends "not many weeks afterward." The farmer, following a bear that has killed one of his sheep, happens on a wild animal's den and finds in the back "the dead bodies, now rapidly decaying, of two small panther cubs."[25]

Predation, Seton and Roberts said, was part of the world. Death was the common lot. "The fact that these stories are true," Seton said in the introduction to *Wild Animals I Have Known*, "is the reason why all are tragic. The life of a wild animal *always has a tragic end*."[26] Predators, then, were possessed of no "hideous or unnatural rage"; they but fulfilled their function. Both authors, albeit with hesitation and ambivalence, followed this logic and found qualities to admire in predators. In one story Roberts's alter ego, the Boy, tries to catch the great horned owl, Hushwing, but not because it did evil. He held no grudge against the great bird "for his slaughter of the harmless hare and grouse, for did not the big marauder show equal zest in the pursuit of mink and weasel, snake and rat?" Even toward that "embodied death, the malignant weasel," he had no antagonism, "making allowance as he did for the inherited bloodlust which drove the murderous little animal to defy all the laws of the wild kindred and kill, kill, kill, for the sheer delight of killing."[27]

Human killing of animals, though, was different. Neither Seton nor Roberts admired sport hunters. Seton wrote that "There was once a wretch who, despairing of other claims to notice, thought to achieve a name by destroying the most beautiful building on earth. This is the mind of the head-hunting sportsman. The nobler the thing that he destroys, the greater the deed, the greater his pleasure, and the greater he considers his claim to fame."[28] In Roberts's stories, man's hunting skills could be unnatural. In one story a hunter, wielding a birchbark horn, calls a bull moose to his death. He is a master woodsman, able to outdo the animals at their own game and to "slay the cunning kindred of the wild by a craft finer than their own."[29] He can mimic with complete fidelity the call of the cow moose, that "noble and splendid call, vital with all the sincerity of response and love and elemental passion." His final call, which overcomes the bull's suspicions, contained all the "yearning of all the mating ardor that had triumphed over insatiable death and kept the wilderness peopled from the first." The story is entitled "A Treason of Nature."[30]

Humans often appear in Seton's stories as destroyers. They kill Wahb's family in *The Biography of a Grizzly*, and Tito's in "Tito: the Story of the Coyote that Learned How." They hunt down the Pacing Mustang and the great ram Krag. Cuddy, the "shiftless loafer," snares the Don Valley partridges in "Redruff." Speaking of Cuddy, Seton said he had changed the name, "as it is the species, rather than the individual, that I wish to expose."[31]

Both authors were ambivalent about that archpredator, the wolf. Roberts accepted much of the standard lore. He believed, for example, that wolves would turn on wounded members of the pack. Only two of his stories were built around wild wolves, and both contain strong moral overtones. "The Passing of the Black Whelps" and "The Invaders" are really one story; Roberts thriftily recycled plot material by changing the locale and some of the action.[32] In both, a bitch runs off with a male wolf and they raise a litter of half-breed whelps. The offspring are worse than either parent—treacherous and implacably hostile to man. In each story the group attacks man and the bitch cannot resist, even in the heat of combat, the man's call. In "The Passing of the Black Whelps," the grown cubs tear their mother apart for this treason, and their wolf father and the man then unite to kill them. In "The Invaders" the bitch survives. One of the hunters reviles her, but the other defends her. She will be better and stauncher, the older one says, for her fling with the wild.

Wolves were much more important to Seton. One of the names he used was "Wolf Thompson"; he often added the sketch of a wolf's paw print to his signature; and among the Woodcraft Indians (his youth group, one of the predecessors of the Boy Scouts) his name was Black Wolf. As an art student in Paris in the early 1890s he often used wolves in his studies of animal anatomy, and the first picture he submitted to the judges for the Grand Salon was "The Sleeping Wolf." It was a powerful, but neutral, study of the animal. The slumbering animal seems to be fading into or emerging from the background, a symbol of the forces of nature. His next project presented another view. Originally entitled "The Triumph of the Wolves," it was based on a newspaper story that appeared while Seton was in Paris. In a remote mountain area, searchers had found the scattered bones of a renowned wolf hunter. Seton's canvas showed the pack at the end of the feast. Blood and bloodstained clothes cover the snow; in the foreground a wolf gnaws at the hunter's skull. In the distance smoke rises from the little cottage where the family waits.[33] At his friends' urging Seton changed the title to "Awaited in Vain," shifting the moral and the center of attention, but it was not nearly enough to divert the storm of criticism that met the painting's public exhibition.[34]

Even in his most famous wolf story, "Lobo," based on his stint as a "wolfer" in New Mexico, Seton was not wholly on the wolf's side. Lobo is a king, but an "outlaw king." He lives by slaughter, is "the hero of a thousand lawless raids," and leads a pack famous for its ferocity and the bloodthirsty slaughter of sheep. Seton traps and kills Lobo, using as bait the body of his mate. The wolf's death converts him, and he lays aside his traps, but there is no hint that any other outcome is possible. Lobo had to die. "It cannot be otherwise," Seton intones. The wolf world is passing, and the wolf with it. Wolves and man are incompatible.[35]

Interpreting Nature Aright

The preservation of the useful and beautiful animal and bird life of the country depends largely upon creating in the young an interest in the life of the woods and fields. If the child mind is fed with stories that are false to nature, the children will go to the haunts of the animal only to meet with disappointment. The result will be disbelief, and the death of interest. The men who misinterpret nature and replace facts with fiction, undo the work of those who in the love of nature interpret it aright.—Theodore Roosevelt (1907)[36]

The new nature stories managed the seemingly incompatible feat of reconciling Darwinian nature "red in tooth and claw" with the Romantic vision of an ordered, purposeful world before whose workings man could stand in awe.[37] Nature was harsh, but its harshness was law. It served the goals of order and purpose. The public acclaimed the stories. With their claims of scientific accuracy, they quickly became the interpretors of nature to a generation of North Americans. But not everyone accepted the new views. One who did not was John Burroughs. In 1903 his essay, "Real and Sham Natural History," set off the "nature-faker" controversy.[38] Burroughs, whose first volume of nature essays had appeared in 1871, was a formidable figure in the world of nature study. He had written almost two dozen nature books and was probably the most widely read naturalist in American history.

He was also an icon. It was Burroughs who would go camping and tramping with that most famous outdoor advocate, Teddy Roosevelt, and write about it. It was Burroughs's picture that appeared on Audubon Christmas cards. It was no small matter when he took to task the authors of the new animal stories.

The first requirement of natural history, Burroughs said, was that it be true. Many recent stories failed this test. He was particularly unhappy with the work of Seton and one of his imitators, the Reverend William J. Long.[39] Seton found far more evidence of animal intelligence than did other naturalists—and found it with astonishing regularity. He pointed to "The Springfield Fox," in which Seton had the fox lure a dog to its death by running across the railroad trestle just before the train came by. This action presumed "that the fox had a watch and a time table about its person." As for the fox's bringing a poisoned chicken head to her cub to deliver it from a life of captivity, that was "pushing the romantic to the absurd."

Long was even worse. His "ridiculous book" (*The School of the Woods*) must, Burroughs thought, have been inspired by Seton's description of a crow college ("Silverspot" in *Wild Animals I Have Known*). The imitator, though, had gone far beyond the original. In Long's book, the animals were educating their young as did humans—reasoning, loving, rewarding, and punishing. If the stories had been presented as pleasant fables, they would have been tolerable; the author's insistence that they were true made them "mock natural history." The pleasure of the tales and the morals inculcated, Burroughs said severely, did not excuse this distortion. Animals did not instruct their young, did not reason, and did not pass along lessons for survival to their offspring.[40]

Seton made his peace with Burroughs. He later claimed, in fact, to have crushed his accuser with his scientific credentials and expertise.[41] Roberts, a minor target, kept silent. Long defended himself at length. He appeared in *Forest and Stream* under the pseudonym "Hermit" (earning editor Grinnell a stern rebuke from Roosevelt for allowing Long to hide under a false name). He appeared again, under his own name, in *North American Review*.[42] Burroughs was wrong, Long said, to judge his stories by the standards of science. Science was concerned with classifying animals; it paid no attention to their behavior, particularly not to their individuality. Nature study was different. It dwelled in an "upper world of appreciation and suggestion, of individuality interpreted by individuality." It was the contact of a sensitive soul with other living creatures. It had a spiritual dimension denied to science.

This put Long and the scientists in different worlds. The controversy might have ended there if Long had not continued to offer as truth accounts of animal behavior that his critics found fantastic. For example, he told of a pair of orioles that could not find a suitable slender branch from which to suspend their nest. They constructed a perfectly triangular scaffolding on the ground (knotting it with reversed half-hitches) and suspended it from a stout branch. In an article on "Animal Surgery," he presented the classic nature-faking tale—the woodcock that made a cast of mud and straw for its broken leg.[43] The eminent scientist William Morton Wheeler asserted that this was impossible. He poured scorn on Long and his followers. Replies ranged from agreement with Wheeler to stout defenses of Long. At the end of Long's final statement (complete with affidavits supporting the orthopedic skills of New England's woodcocks), the editors of *Science* said they hoped "the discussion would not be carried further."[44]

Vain hope, for the issue was not Long's credibility but the extent to which animal behavior was like our own. How did animals learn, if at all? In *Ways of Nature* (1905) and in an article in *Harper's*, "Do Animals Think?" Burroughs came down on the side of instinct. We tend, he said, to talk of animals in familiar terms, and to interpret their movements and actions by analogy with our own mental processes. However, "all modern comparative psychologists account for all their actions without attributing to them any of the higher human facilities." He cited Edward Thorndike, one of the pioneers of experimental animal psychology, on animals' lack of self-consciousness. They lived "wholly immersed in the unconscious, involuntary nature out of which we rise, and above which our higher lives go on," on "the plane of sense . . . only now and then just breaking for a moment into the higher plane. In the world of sense man is immersed also—this is his start and foundation, but he rises into the plane of spirit, and here lives his proper life. He is emancipated from sense in a way the beasts are not."[45]

Four months later *Harper's* printed a second article entitled "Do Animals Think?" This one was by Peter Rabbit (Long behind a set of false whiskers, or false fur). Three months later "The Question of Animal Reason" appeared under Long's own name.[46] Animals, Long said, were, like men, capable of thought; like men, they did not always think. We know that animals can reason because we can see them doing it. We can infer their mental processes by watching what they do and thinking what we would be thinking under those circumstances. He did not expand on this—though that extrapolation was the crux of the whole debate—except by appealing to the unity of nature. There was a single set of laws for the mind, he said, just as there was a single set of laws for chemistry and for physics.

Another article by Peter Rabbit, "Animal Immortality," expanded on this theme. People, the rabbit said, should look at their own situation, see what hope of eternal life they had, and apply that to animals. Animals, like people, had an inborn sense of right and wrong and a personality. Were their lives and personalities less deserving than men's? Animal lives, more than human, were incomplete, cut off, and this would be remembered by "any great Love or Wisdom overlooking the Universe." Animals were sensitive to supernatural influences; when ghosts appeared they were the first to notice. Finally, as an animal sank toward death, it thought it was going to sleep as it always had and that it would awake: "Life and Nature, which have spoken truth to the animal all its life, surely will not deceive it in the final moment."[47]

Long and Burroughs appealed to different psychological theories, but the argument was really about the proper interpretation of nature and the place of nature interpretation in the life of the country. This was most apparent when Theodore Roosevelt joined the fray. He had been fuming for years, in private, about bad natural history. He was particularly upset about Long, whom he dismissed (in a letter to Burroughs in 1903) as a "ridiculous creature." He had kept silent because he did not think it fair for the president to chastise a private citizen for a private matter, but in 1907 he lost his temper. He blasted the "nature-fakers" in an interview. The stories were not, he declared, just wrong; they were harmful. It was upon the interest of the young that the "preservation of the useful and beautiful animal life of the country" depended. These people, creating false expectations and ideas, destroyed the genuine interest. He condemned Long as "perhaps the worst" of the group. He scoffed at the school for young caribou, where they learned how to live in the wild, and at Long's tale of the mighty northern wolf whose fangs could reach a caribou's heart in a single snap. Ridiculous, snapped the president.[48]

Long replied in a full-page interview in the *New York Times*, under the title "I Propose to Smoke Roosevelt Out—Dr. Long." He stuck by his work and even took the offensive. He wanted, he said, to "bring a new spirit of gentleness and sympathy into our study of animal life." That was "perhaps the chief reason for [Roosevelt's] antagonism," for the president was a bloodthirsty slaughterer of wildlife. Long concluded that if he had done something to counter the spirit of the president's approach to wildlife, he would be content.[49]

By then less reverent spirits had found yet another approach to the subject. *Outing* printed in its fiction section "A Fact or a Fake submitted to Mr. John Burroughs." It told of a dog who, deprived of her pups, had gathered a group of dolls as substitutes.[50] On the page preceding Long's defense, the *New York Times* printed a piece by the popular humorist Peter Finley Dunne, "Mr. Dooley on the Call of the Wild." The perspective from Mr. Dooley's tavern was somewhat different than from the drawing room. Mr. Hennessey, the bartender's "literary advisor," had apprised him of the new nature stories. Mr. Dooley had hastened to enlighten himself. The result was a hilarious putdown of everyone. "Tiddy" was outraged that a nature writer had a guinea pig "killing a moose by biting it in the ear. Now it is notoryous to any lover iv th' wilds," Mr. Dooley reported the president saying, "any man with a fondness f'r these monarchs iv the forests, that no moose can be kilt be a wound in th' ear. I have shot a thousand in th' ear with no bad effects beyon possible makin' thim hard iv hearing."[51]

LAUGHTER may have swept away the controversy; in any event, it died soon thereafter, despite a second blast from the president. But the stories did not stop. People wanted to find in nature—the nature into which Darwinian evolution had thrown them—order, harmony, and human virtue. The new nature fiction allowed them to do that. It interpreted science in a comfortable way, even as it entertained. Just how deeply the new ideas reached can be seen in the changed place of predation and predators in children's nature stories. It was one thing to portray the struggle of life for adults, and quite another to do it for children. Yet this was the message of one of the most popular of all nature writers at the turn of the century, Thornton W. Burgess. He wrote an immense amount, much of it in small chapters that could be printed serially in daily papers before being made into books. He aimed at elementary-school students. In his stories, jolly round red Mr. Sun comes up over the Green Meadow and Old Mother West

Wind brings the Merry Little Breezes down to play with Peter Rabbit, Johnny Chuck, and Reddy Fox. The animals speak, wear clothes, and behave much like their intended audience—small boys. But there is also natural history here, and Burgess did not flinch from the implications. We all know that Reddy lives on mice, frogs, rabbits, and young chucks. When the wishing stone allows Farmer Brown's boy Tommy to experience life as a fox, he too lives that way. On his first day as a fox, in fact, he cooperates with Reddy to fool Mrs. Chuck and get one of the little chucks for a meal. On becoming a boy again (as he must at the end of the chapter), Tommy has a new perspective on Reddy. "He may kill a lot of innocent little creatures," Tommy thinks, "but he has to live, and it's no more than men do (he was thinking of the chicken dinner he would have that day)."[52]

Predation is a common theme in Burgess's fiction. Hooty the owl is a terror to all the little animals in the Big Green Woods, and so are the hawks. When Peter Rabbit goes to Old Mother Nature to learn about the birds and animals, she includes those who live on their fellows, and even the most fearsome gets his due. Every creature dreads "Howler the Wolf," but wolves are "the best of parents, and the little wolves are carefully trained in all that a wolf should know. Always the hand of man has been against them, and this fact has developed their wits and cunning to a wonderful degree."[53]

Burgess also relied on science. He wrote to interest children and to increase their sympathy with the wild creatures, but he built his stories around natural history, and he took pains to be accurate. Occasionally the science showed through. At one point Tommy Brown lectures a man hunting on his father's land about shooting hawks. The birds, he points out, eat many mice and are very valuable to the farmer. Tommy has clearly been improving his idle hours after jolly round red Mr. Sun had gone down behind the Purple Hills: he has read A. K. Fisher's *Hawks and Owls of the United States*, published by our friends in the United States Bureau of Biological Survey.[54]

Burgess's science was more than a matter of assimilating Darwinism. He relied on something new in America: an organized, self-conscious community of scientists. The development of this group, which was centered in government bureaus, academic departments, and research museums, would have important consequences for our ideas about wildlife and wildlife policy. It produced information and experts. It would provide expert guidance for wildlife policy. Its members would take active roles in public movements to preserve wildlife, and its research would provide the backbone for public protests against the destruction of nature. The development of ecology and its growth within the scientific community were among the most important advances in wildlife work in early twentieth-century America.

CHAPTER THREE

Ideas and Organizations
Structures for a Wildlife
Policy

In the decades around the turn of the century, Americans laid the institutional and intellectual foundations for a wildlife policy. They established agencies and programs and passed laws that provided a basis for government protection of wildlife. The science of biology was also changing. Its ideas and the institutional and social arrangements that generated knowledge were developing rapidly. By the 1920s the internal supports for ecology and for the application of ecology to wildlife work were in place. A federal wildlife agency, the Bureau of Biological Survey, was established (later it would become the Fish and Wildlife Service). It had definite responsibilities, a research program, and the potential to coordinate a national program—all of which biologists were now able to support. They were no longer a collection of amateurs: they were a self-conscious group with a definite program and becoming increasingly specialized. And those specializing in the science of ecology had begun to find their own concepts and research techniques.

Little of the effort that went into building institutions or developing ideas was consciously directed toward the question of wildlife policy. There was at this point no real "policy," only a collection of laws guided by "common sense." The changes were, however, a vital part of the chain that led to modern attitudes. The first protests against wiping out the "useless" animals would come from the institutions that supported the science of ecology, the institutions that would also be the basis for the modern structure of wildlife agencies. The public became interested in "vermin" in large part through learning about the creatures from popular presentations of scientific research done in agencies and academic departments; and, later, when conservationists demanded that wolves and other such animals be saved, they would turn to these agencies and their science for direction.

34

A Federal Presence

The federal government almost stumbled into wildlife work. In 1884 the American Ornithological Union, a professional association of scientists studying birds, asked Congress to set up a new office to study the economic relations of birds and mammals. The next year, possibly influenced by the lobbying of Spencer Baird, secretary of the Smithsonian Institution, Congress appropriated funds for an Office of Economic Ornithology and Mammalogy. It was a modest beginning for what would become, as part of the Department of the Interior, the U. S. Fish and Wildlife Service: one naturalist and a clerk in an office attached to the Division of Entomology of the Department of Agriculture.

The office showed no immediate promise of becoming an important or particularly promising scientific institution. Indeed, it seemed at first to be mired in the past. Its chief, C. Hart Merriam, was one of the "last of the naturalists," the medically trained or self-trained students of nature who dominated the study of field biology into the early twentieth century.[1] Born in 1855 to a well-to-do Massachusetts family, Merriam was educated by governesses until he was fourteen, and then sent to private school. When he was sixteen his father introduced him to Baird, then a curator at the Smithsonian. This connection got him a position on the Hayden expedition to Yellowstone in 1871. He then served an apprenticeship at the Smithsonian, where he met virtually all of the country's important naturalists. He became a physician but continued to work in natural history, spending his summers at the Woods Hole Laboratory of the U.S. Fish Commission. One suspects that it was with a certain sense of relief that he abandoned medical practice to take up his new appointment as head of the Office of Economic Ornithology and Mammalogy.

Merriam was not, in some ways, a good choice for the job. Though he was ambitious for his organization and a hard worker, he had little of the tact and political savvy required in dealing with congressmen. They were interested in what the Office (which Merriam soon renamed the Biological Survey) could do for America's farmers. But Merriam was interested in natural history for its own sake. He was so little inclined to bend to Congressional demands for "practical" research that he spent most of his tenure (he continued as chief to 1910) in political hot water. On one occasion Congress stopped the agency's

35

appropriations, and only the intervention of his friend President Theodore Roosevelt saved the day.[2]

Merriam was just as out of touch with the biology of his time. Even as he began his tenure natural history was dying or changing. Graduates of specialized programs were replacing people who learned their craft through apprenticeship, informal study, or taxidermy. Laboratory research replaced fieldwork, studies of parts of an organism became more prominent than work on the whole. Taxonomy and the distribution of species, the principal problems in Merriam's day, were losing their appeal.[3] Merriam fought these trends. He grumbled about "microscope biology" and called for more fieldwork. As head of the Survey he relied on people without formal training. Of the dozen or so men he hired and promoted—who dominated the Survey for decades—only two had Ph.D.'s, and two M.A.'s; five had no degree at all. They were, much more than the scientific community of the time, old-line naturalists.[4]

Despite his prejudices, Merriam wound up encouraging scientific work not only in the agency but in academia as well. He almost had to; academic biologists and Survey workers needed each other. The Survey had an annual appropriation (however small) and a mandate to do scientific research on birds and mammals. It could undertake long-term studies that were impossible for most institutions.[5] To do this it needed trained scientists. Academic biologists, on the other hand, wanted and needed the experience and the knowledge they could get on Survey expeditions. The need for access to specimen collections was another bond. The Survey had its own specimens and access to the Smithsonian's, but it also relied on the collections of the Museum of Vertebrate Zoology at the University of California, Berkeley, and the American Museum of Natural History in New York, and local museums. Specimens, papers, and people went back and forth, and work around the campfire and the specimen case converted professional associations into personal friendships.[6]

The Survey also helped the specialization of biology, occasionally quite directly. In 1919 a group of Survey workers organized the American Society of Mammalogists. They elected Merriam its president. Then and later federal workers were an important part of the society and filled many of its offices.[7] The agency became a clearinghouse for scientific information and policy. It published collections of state laws, surveyed changes in state practice, and worked against what

it considered poor policy—for example bounties, which it characterized as useless, expensive, and open to fraud. In the 1930s, it worked
to organize game management as a science. It provided an abstracting
service for wildlife research papers. Finally, it directed cooperative research programs with the states and was the nucleus for what became
the Wildlife Society, a professional group of game managers.[8]

While the Survey shaped science, its position in the government
shaped it. As Congress passed wildlife laws, it charged the Survey with
enforcing them. Administrative responsibilities meant regular appropriations, but they also forced Survey biologists into new roles. They
had to meet the public, enforce laws, and negotiate with states or (in
the case of duck-hunting regulations) with a foreign government.
Agency officers became administrators and law enforcement officers,
and the agency something more, and less, than a scientific institution.[9]

The Lacey Act of 1900, the agency's first law enforcement responsibility, started the watchdog process. To enforce the provisions barring interstate shipment of game killed in violation of state laws, the
Survey had to build a law enforcement division. Since Congress did
not provide funding for a federal wildlife police force, the Survey had
to cooperate with the states, conservation and bird protection associations, furriers, and milliners. It had, in short, to build a network of
alliances and relationships across a wide range of interests. Another
clause gave the Secretary of Agriculture authority to ban the importation of pests, and responsibility for managing already established
pests such as starlings. That forced the agency to develop similar relationships with the Customs Service, sportsmen, and importation societies.

Duck hunting added further dimensions to the Survey's work. For
years a coalition of sportsmen and nature lovers had sought federal
aid in protecting migrating waterfowl. States were powerless, for the
annual migrations between the tundra of Arctic Canada and the Gulf
Coast of America cut across dozens of states and provinces. The federal government, however, had no explicit authority to manage wildlife. In 1913 Congress found a way around this. The Weeks-McLean
Act gave the Secretary of Agriculture authority to regulate duck and
goose hunting seasons on the grounds that the migration of waterfowl
across state boundaries constituted interstate commerce.[10] Not surprisingly, there were legal challenges to regulations issued under this act.
To avoid adverse rulings, the United States and Canada signed a

treaty, negotiated by the Survey and Canadian wildlife officers. The act Congress passed—the Migratory Bird Treaty Act of 1918—gave the regulations a firmer legal foundation: treaties took precedence over state law. It also gave the Survey international responsibilities. The two countries had to agree each year on seasons, bag limits, and protected species.[11]

To do that the Survey had to develop census techniques, chart migrations, and learn how to predict the level of the waterfowl population. This required a progam of applied ecological research. The creation of wildlife refuges under the act called for additional research—on food, cover, and disease. The Survey also had to negotiate with the Canadians and, when working with wildlife refuges, with local residents angry at the closing of choice hunting areas or at the loss of crops to the migrating flocks.

The Survey's responsibility for predator and rodent control affected it even more. Like other areas of wildlife work, this was a case of the federal government moving into what had traditionally been a matter of state action. The difference was that here there was a demand for federal help and no opposition. As early as the 1880s, ranchers on the plains and in the Great Basin were appealing for help in controlling rodents and predatory mammals. States set agricultural experiment stations to work on poisons and poisoning formulas. They passed bounty laws and in a few cases handed out free poison. Even taken together these did not solve the problem. By the turn of the century there were calls for compulsory eradication laws to control ground squirrels and prairie dogs, with the state taking over the work for any landowner who refused and charging the cost to his taxes.[12]

As a major landowner in the West, the federal government came under pressure to help or at least to do its share. In response, the Forest Service in 1905 began hiring trappers to kill wolves in the national forests. It also arranged for Survey biologist Vernon Bailey to find out where the wolves lived and bred.[13] The Survey took an interest, in large part because Merriam saw in the issue a chance to reply to critics who claimed the agency was doing nothing practical. He spoke about the menace of coyotes and wolves, which "levy a heavy tax on stock . . . a loss aggregating several millions of dollars annually."[14] The Survey hired David Lantz from Kansas to work on formulas and distribution equipment and to advise ranchers on the best ways to get rid of "noxious animals." Around 1912 it began demonstration projects and field tests.[15]

Congress soon provided formal responsibility. In 1914 it funded demonstration projects, and the next year it appropriated $125,000 for a continuing program of predator and rodent control, to include the Forest Service's trap lines.[16] It justified the program as a contribution to the war effort and a way to combat a rabies epidemic then raging in the Great Basin. The crucial factor, though, was the livestock industry. Western congressmen saw federal predator control as a cheap way to serve their constituents. Their colleagues did not object—they had their own pet projects—and, in 1915, the idea of killing off "varmints" offended no one.

There was, the Survey soon discovered, more demand for predator control than money to do it, so the agency began shifting the cost to the ranchers. After 1918 almost all work was done under cooperative agreements. State or local authorities, grazing districts, or cattlemen's associations would seek help. The Survey would investigate (often a formality), and if control indeed seemed necessary, it would provide the hunter and equipment. The "cooperators" (as they were called) paid the bill, usually by a tax on livestock in the affected area. By the mid-1920s these funds were a quarter of the Survey's budget, and the ranchers were the main support of the new Division of Predator and Rodent Control (PARC).[17] The scheme had something for everyone. Congressmen had the satisfaction of having helped their constituents. The Survey got a source of outside funds, a continuing responsibility that justified its existence, and a set of clients who provided political support at budget time. The ranchers got better predator control. Government agents, unlike bounty hunters, did not quit when prices were low or wolves scarce. They could be, and were, assigned for months at a stretch to get rid of some notorious "outlaw."[18]

In the Dakotas, Wyoming, Colorado, New Mexico, and Arizona, the last resident breeding wolf packs vanished by the mid-1920s. Wolves survived only as remnant populations in wild country where no stock was raised. The PARC, with its own mission, clients, and funds, began to change the agency. He who pays the piper calls the tune, and when the ranchers began paying they began calling. Scientific studies and the conservation of wildlife became less important than return on money spent and a high kill of "varmints." W. L. McAtee was only one of the old-time employees who came to feel that the old agency he had

39

joined in 1904 was dead, that science had been discarded in favor of wildlife destruction.[19] Sometime in the 1920s the derisive term "go-pher-choker" began to circulate within the agency—a visible indication of ill-feeling.

The split was not just within the Survey; the scientists themselves were of two minds. Most of them were deeply interested in animals (why else work for the Survey?). They were also creatures of their society, committed to science, rationality, efficiency, and modern methods. This led to some curious stands. Vernon Bailey, one of Merriam's original crew and chief field naturalist for the agency from 1887 to 1937, invented several humane animal traps and was deeply committed to humane treatment of animals. In his papers are photos´of trapped wolves with captions like these: "A Big Gray Wolf, in the snow. Caught and Held in two no. 2 Steel Traps. Feet Frozen but no less Painful." "Yes, he killed Cattle, to Eat, But, Did he Deserve This?"[20] But Bailey also condemned wolves as "cattle-killers" and said that the species could not be tolerated near "stocks" of game animals. He scoffed at a "balance of nature" and called on humans to manage all land—and that meant an end to predators.[21]

Stanley Paul Young, a veteran "gopher-choker," PARC worker from 1919 to 1937, and chief of the division from 1937 to 1939, was equally divided. In the late 1960s, writing of the PARC campaign forty years earlier, he called the last western wolves "criminals" and "outlaws" who had written "their own death warrants in killing, torture, blood lust, almost fiendish cruelty."[22] He collected and kept, apparently without thought, photographic jokes: one picture showed a dozen dead coyotes draped like deer trophies over a car with one propped up at the steering wheel as if driving.[23] At the same time, he devoted years to a study of the taxonomy and prehistoric distribution of the wolf in North America. With E. A. Goldman, another PARC officer, he wrote the definitive scientific treatment of wolves for his generation, *The Wolves of North America* (1944). With H.H.T. Jackson (known in the Survey as "Alphabet" Jackson) he wrote a similar work about the coyote, *The Clever Coyote* (1951). Both books offer many admiring comments about their subjects. Even in an article, "The War on the Wolf" (1942), he was able to say that "Where not in conflict with human interests, wolves may well be left alone. They form one of the most interesting groups of all mammals, and should be permitted to have a place in North American fauna."[24]

Science and the Scientists

But what was that place and where would it be? Today science answers these questions, or at least sets the terms of debate, but at that time it could not. Like the Survey, which grew from a clerk and a naturalist into a large, established agency, biology—its scientific base—also had to develop. It did so as part of the specialization of American science in the late nineteenth century, shaped by its growth within the academic structure of the university. In 1882 the American Association for the Advancement of Science split its Natural History section into nine new divisions. Biology was one of them, and it split almost immediately into new groups—botany, mammalogy, and entomology to start. It was also changing its focus, moving from field observations to laboratory experiments and from a study of the whole organism to a concentration on parts and processes.[25] While this was going on, the new land-grant colleges, with their associated experiment stations, were fostering applied biological disciplines.

Ecology developed within this milieu of specialized, academic studies, but it did so in its own fashion. It drew on both pure and applied research. Its roots were in plant and animal geography, limnology (the study of freshwater organisms), and oceanography and projects such as the control of the gypsy moth and the cotton boll weevil. It developed in places as diverse as the University of Chicago (a "pure"-research institution) and the University of Nebraska (one of the "cow colleges" founded under the Morrill Act to provide instruction in "agriculture and the mechanic arts"). It was heavily influenced by Darwinian ideas and by the older traditions of natural history and nature study.[26]

The career of Stephen A. Forbes, one of America's pioneer ecologists, was typical in its variety. Born in 1844 the son of an Illinois farmer, Forbes had a sketchy education: service in the Illinois cavalry during the Civil War and attendance at Rush Medical College in Chicago, where he failed to take a degree. He became a scientist by apprenticeship, and in that open era advanced quickly. He began publishing scientific articles in 1870 and two years later became head of the Illinois Natural History Society Museum. He was also appointed professor of zoology and entomology at the state university, and state entomologist. Besides doing his academic work, he ran one of the earliest continuing natural-history agencies, the Illinois Natural History

Survey, which trained and gave experience to a variety of scientists—including some of the ecologists of the next generation.

Forbes's research was as varied as his responsibilities. He did pioneer work in limnology. His *The Lake as Microcosm* (1887) was a major advance in our understanding of the distribution and abundance of freshwater fish and a defense of the balance of nature in Darwinian terms.[27] He studied the factors influencing insect abundance and the mechanisms of population control in insects. He was one of the first to suggest that insect diseases might be used to control pest species. He was a pioneer in the analysis of stomach contents of birds and used the results to determine not only their diet but their economic importance. The work, besides its application to agriculture, threw light on population dynamics.[28]

Forbes was interested in Darwinism, but he and his contemporaries also spent much of their time on what we would now consider "ecological" questions—particularly population dynamics and the mechanism of population control. Circumstances all but forced these problems on scientists, particularly those in the U.S. Department of Agriculture (one of the leading employers of scientists at that time), the land-grant colleges and experiment stations, and the state agricultural agencies. Steamship traffic and deliberate introductions brought into the United States and other countries many organisms that, lacking established natural enemies or other controls, multiplied at enormous rates. In 1897 the head of the USDA's Bureau of Entomology, Leland O. Howard, estimated that a third of the major crop pests in the country had come from abroad. Lacking effective, inexpensive insecticides, scientists were forced to try other kinds of control: they imported insects that ate pests; and they tried to modify farm practices to eliminate the food, hibernation spots, or other things the pest required.[29] To do this they needed knowledge of the ecological requirements of the pests and their enemies.

Deliberate importations brought another set of problems. Sportsmen tried to get birds and animals from Europe, Asia, and Africa to replace "shot-out" native species or to provide extra gunning. Homesick Europeans and interested Americans brought in songbirds and cage birds, ranging from skylarks and nightingales to starlings.[30] Introductions were a worldwide phenomenon. So were the irruptions—sudden large increases in wildlife populations— that followed. By the 1890s a Biological Survey scientist could pass over the case of the European rabbit in Australia with the observation that "the results of the experiment are so well known that anything more than a brief reference to them is unnecessary."[31] In the United States during the same period that the cotton boll weevil was sweeping through the South, the gypsy moth was causing major problems in New England, and the starling was making a nuisance of itself along the East Coast.[32]

Irruptions concentrated biologists' attention on the mechanisms that normally kept them in check. It was clear that *something* did (the sea, after all, was not full of oysters, despite the millions of eggs each of the creatures produced). It was equally apparent that the mechanism could go awry. Beyond that, things were dark. The common explanations, the "balance of nature" or the "economy of nature," were, like "instinct," terms without content.[33] People often appealed to predation. The carnivores, in this scheme, lived off the herbivores, which bred frantically to keep up their numbers. This notion was just a theory, and attempts to apply it showed that there was much more involved. After a decade spent importing European and Japanese insects that preyed on the gypsy moth, entomologists found it might require populations of some fifty species of its natural enemies, eating every life phase from egg to adult, to keep the moth down. They also realized that the work would require detailed study of the life history of all these species. The scientific results of these projects were important for animal ecology; but they did more to show the difficulties in formulating a comprehensive ecological theory than in actually building such a concept.[34]

It was a long way from insects to wolves and deer or to coyotes and rodents, and no one immediately made the leap. There were good reasons not to. The ecology of larger predators and their prey presented technical problems that would discourage researchers for the next sixty years. Wolves ranged widely and swiftly over enormous areas of forested wilderness land. Even counting them was a formidable task; discovering their relations to other species and to the environment was even harder. Work on other species, however, was important. It provided examples for mammalogists working on wolves and coyotes. It suggested a way of thinking and a set of questions that might in time be applied to these elusive creatures. Moreover, it forced field biologists to think in terms of relationships and effects, not taxonomic criteria.

Plant ecology developed before animal ecology, possibly because plants did not run off and hide as the scientist approached.[35] The factors governing the distribution of plants across the earth and the associations among them were old problems and among the first that scientists studied. Study in this area led to the first ecological concepts that were made popular with the backing of science. Merriam, surprisingly, led the way. In 1891, on the basis of the Survey's study of San Francisco Peak in New Mexico, he proposed dividing the North American continent into a set of "life zones," each with characteristic vegetation and associated animal species. Temperature, he said, set the boundaries.[36] Natural historians had been making such schemes for a century, but Merriam's caught the public eye. It was simple and it had the authority of the government printed on every map. It was also available just as the wave of "nature education" washed over the public schools. Generations of schoolchildren studied idealized diagrams of a mountain peak with zones running from Arctic at the top to tropical at the bottom. They learned that altitude and distance north could be linked. Museum exhibits, popular handbooks, and magazine articles spread Merriam's ideas into the culture.[37]

Even as Merriam was working out his scheme, Frederick Clements, a student in Charles Bessey's botanical seminar at the University of Nebraska, was working out the concept that would dominate American plant ecology for the next fifty years—the climax theory of plant succession. The contrast between Merriam and Clements shows how much American science had changed. Merriam had a medical degree and had become a scientist by apprenticeship. His ecological ideas were in the tradition of early nineteenth-century biogeography, his research methods qualitative. Clements, on the other hand, had been professionally trained as an ecologist in the first center of ecological studies in the country, Bessey's botanical seminar, which was organized to support professional research projects. Its objectives were defined, as much as such things can be, by questions arising within the discipline.[38]

Clements's work was grounded in quantitative and statistical methods. He used sample plots and counted the numbers of individuals. Looking at the prairie in this new, statistical way, he found an ordered sequence of changes. Each area, beginning as bare ground, was dominated by a succession of plant associations. Each, which formed a "superorganism," prepared the ground for the next, culminating in the "climax" stage, an association that was self-reproducing and determined by the climate.

These ideas had their critics and were never popular with European ecologists, but they dominated American ecology and American education into the 1950s. Along with Merriam's life zones, Americans learned how bare areas were seeded with pioneer species, how these yielded to others, and so on, until (at least in the East) there was a mature forest. Grade-school texts, nature magazines, and even the Boy Scout handbook presented this picture of a nature moving inexorably toward a steady state.[39]

At the University of Chicago, Henry Chandler Cowles's graduate seminar did for animal ecology much what Bessey's seminar at Nebraska did for plant studies. Like Bessey's, Cowles's group began with succession. Inspired by the work of the Danish scientist Eugene Warming (whose work, in German translation, became available in America in 1896), Cowles investigated the succession of plants in the Indiana sand dunes. Victor Shelford, one of his students, repeated this work for animals. He charted the distribution of species of tiger beetles in the dunes and their relationship to the plants. The group moved beyond specialized studies to larger projects. Charles C. Adams, a Cowles student who had worked for Forbes's Illinois Natural History Survey, wrote a *Guide to the Study of Animal Ecology* in 1913. A survey of the literature and a manual for fieldwork, it was the first American attempt to build the scattered observations on the distribution and relationships of animals into a discipline.[40] Four years later, Shelford classified the animal communities of North America. It was, he recognized, not entirely satisfactory, for it used the associations established for plant communities. Still, it was an acceptable stop-gap.[41]

Professional organization soon followed. In 1914 Shelford, Cowles, and a Nebraska ecologist suggested forming a society, and the next year twenty-two scientists met in Philadelphia and founded the Ecological Society of America, with Victor Shelford as its president. In 1920 the society began publishing *Ecology*, the first American journal in the field.[42] Ecology was still, however, something less than well defined. It seemed at times to resemble a set of ideas in search of an organizing principle. There were very basic conceptual problems. What was the proper unit of study—the species or the habitat? How should study areas be bounded? Aquatic environments had definite borders—which helped limnology develop—but terrestrial communi-

45

ties were often hard cases. Ecology, already divided into plant and animal ecology, split further into autoecology (study of a single species or population) and synecology (the study of a particular area). Some ecologists abandoned field populations for laboratory studies or mathematical models.[43] There were brave but futile attempts to bring order out of chaos by terminology, leading to the derisive definition of ecology as that part of biology given over entirely to nomenclature.

Ecologists were not even clear where the field fit into biology. Shelford defined animal ecology as a branch of general physiology dealing with the entire organism that also considered the organism with reference to its normal environment.[44] That same year Adams said that ecology was concerned with the responses of organisms to their complete environment. It was not, in his view, a branch of general physiology, but one of the fundamental divisions of biology.[45] Others made more expansive claims, calling on ecologists to incorporate sociology and the other sciences dealing with human interactions. The science was developing, but in organization as well as in ideas it had not yet found its place.

But by 1920 a scientific establishment existed in which ecology could find a place. Academic departments, research collections, and government agencies concerned with saving wildlife were well established. There was also a strong, if undefined, public demand for what the ecologists could supply. Americans wanted to save wildlife for sport and appreciation. They needed information for that—ecological information. The determination to save nature in our industrial society would support ecological research, but even as ecology was maturing, some attitudes were already beginning to change. The heretofore unthinkable was developing—a demand for the preservation of the wolf, the coyote, and other "worthless" wildlife.

Interlude

Values and Varmints, I

Before knowledge came commitment. In the 1920s mammalogists, who were professionally concerned with wildlife, made the first organized attempts to change government policy toward predatory animals. Their target was the Bureau of Biological Survey's poisoning program. The debate between the Survey and the mammalogists was not a direct precursor to the environmental debates of the 1970s, and it was carried on under different conditions. The debate was not public; it involved only the professional and scientific communities, and it was not buttressed by the kinds of scientific studies that are used now in debating policy. The exchange was a good bench mark, however—a vantage point from which to observe a changing science and the public's reaction. It was a taste of what was to come.

CHAPTER FOUR

Worthless Wildlife

In 1928 Paul G. Redington, chief of the Biological Survey, declared
that "We face the opposition . . . of those who want to see the mountain lion, the wolf, the coyote, and the bobcat perpetuated as part of
the wildlife of the country."[1] He exaggerated; the "opposition" was
only protesting the Biological Survey's extensive use of poison to kill
coyotes. Still, its very existence indicated a fundamental change. It was
one thing to save the buffalo and the plume birds. They were part of
America's national heritage or aesthetic delights. The wolf, coyote,
and bobcat were the symbols—and the substance—of the savage wilderness Americans had sought to conquer. To save them, even to oppose their elimination, was to repudiate the pioneer past. The existence of an opposition to predator control indicated that at least some
people were attaching new values to varmints.[2]

By the time federal predator control work started in 1915, private
hunting and bounty hunting had reduced the larger predators—wolf,
bear, and mountain lion—to remnant populations, and the Survey
quickly got rid of the remaining "outlaw" wolves. By the mid-1920s
only one target numerous enough to justify a full-scale control program remained—the coyote.[3] Neglected when its larger relative was
abundant, it now became an "archpredator" and a threat to western
stockraising.[4] Woolgrowers were most concerned; their charges were
just the right size for coyotes. Their desire for a varmint-free range
and the Survey's need for clients meshed, and the PARC, with the
enthusiastic support of the National Woolgrowers' Association, began
an all-out war against the "brush wolf." Until 1929 the agency described its annual progress under the heading "extermination." The
term was dropped but the objective remained. The PARC sought to
save stock by killing as many coyotes as it could. The favored method,
because of its economy, was poison. The Survey began with "drop
baits"—single pieces of fat or meat poisoned with strychnine—and

went on to "bait stations," a quarter or half a horse laced with the same poison.

After World War I the PARC expanded its wartime program, and by the mid-1920s was killing 35,000 coyotes a year.[5] As the program expanded, mammalogists became concerned. Westerners took the lead. They were, of course, closest to the area, and many had grown up in the country. They had seen it when it was relatively untouched and were appalled by the rapidity with which development was affecting even remote areas. Like the anthropologists who avidly collected every scrap of Indian lore, they stood at the end of a world, and, like the anthropologists, they knew it and fought against it. They could not deny all change—did not want to—but they did want to stop what they saw as useless destruction of wildlife.[6]

Joseph Grinnell, director of the Museum of Vertebrate Zoology at the University of California, Berkeley, was their leader. The head of the premier institution of its kind on the West Coast, he was also firmly committed to the preservation of native wildlife. To shield his institution from angry state legislators, he let his assistants Joseph Dixon and E. Raymond Hall make the public statements. He was in some ways a peculiar man to lead a crusade for wildlife. He was not notably Romantic nor was he an advocate of humane treatment of wild animals. In fact, Grinnell seemed the complete museum biologist, more interested in dead animals than in live ones. He supervised the museum, which he directed from its founding in 1908 until his death in 1939, in minute detail, choosing even the tags and the ink for specimen labels. He insisted on complete, cross-referenced correspondence files (to the delight of historians), and he ran the museum with an iron hand. Deviations from his standards earned the culprit a written note, which was to be returned when the condition had been corrected; Grinnell confirmed the change, initialed the note—and then put it in the person's file. Once, when Grinnell telegrammed from the field that some specimens he was sending in be prepared immediately, Raymond Hall pleaded that the diligent museum staff should be allowed to take Christmas Day off. Besides, Hall added, the freight office was closed.[7]

For all his pedantry, though, Grinnell loved nature—a love nurtured by his early life and only reinforced by science. He had been born in the Indian Territory (what is now Oklahoma) in 1877 and grew up there and in California. Since his youth he had been an am-

49

ateur collector of birds (he went to Alaska with the gold seekers and paid more attention to birds than pay dirt), and his field trips for the museum kept him out of doors. His training made him conscious of how people affected the land: he could see changes in distribution of rodents, and he watched wildflowers vanish from mountain pastures, the victims of overgrazing. Even before the predator-poisoning issue arose he had begun to speak out for wildlife and wild lands.[8]

Though centered in the West, the dispute over predator control affected the entire professional community. With one exception, every president of the American Society of Mammalogists between 1919 and 1954 (when the Society began electing people too young to have been involved) had been either a Survey employee in the 1920s or a protester. Several stiff exchanges of letters took place among formerly cordial scientists, and more than one broken friendship resulted. Ira Gabrielson, a Survey biologist, resigned in protest from the American Society of Mammalogists after the Society censured the Survey's poisoning program.[9]

Discontent became open opposition at the 1924 meeting of the American Society of Mammalogists, where several members debated predator-control policy with two of the Survey's biologists, E. A. Goldman and W. B. Bell.[10] Critics maintained that the Survey was not controlling predators but exterminating them. Valuable material, they complained, was being lost to science. Local populations were being wiped out and some subspecies might already be extinct, making it impossible to reconstruct their place in the area or their relation to other populations of the same species. There were conflicting economic considerations as well. Dead, predators furnished fur, which was an important source of money in some parts of the West. Alive, predators controlled or curbed rodents that competed with cattle and sheep for grass. Finally, the mammalogists pleaded, the animals might have other, as yet undiscovered, values. C. C. Adams cited the recent discovery of insulin in sharks' liver to show what good might come from "vermin."[11]

Goldman, who bore the burden of defending the Survey, said that scientists need not be alarmed. Predator control might cause "local extermination," but the Survey did not intend to kill off whole species, and was not doing so. The smaller predators, particularly the coyote, were in no danger at all, he claimed. Large predators might be lost from the United States, but they would still survive in Canada and

Mexico. His main point, though, was that poisoning was necessary: "Large predatory mammals, destructive to livestock and game, no longer have a place in our advancing civilization." They were an unacceptable drain on resources. In any event, the Survey was not responsible for the policy of killing off predators; ranchers had been doing that for years, and the Survey was only continuing what they had done, more carefully and on a more rational basis.[12]

Goldman's rhetoric did not satisfy everyone. The Society of Mammalogists directed its president to appoint a committee to report on the Survey's predator-control work. The group selected was as balanced as it could have been: two Survey biologists, Vernon Bailey and E. A. Goldman; and three academics, Joseph Dixon of the Museum of Vertebrate Zoology, C. C. Adams of the New York State College of Forestry and the Roosevelt Wildlife Station, and Edmund Heller (chairman), director of the Milwaukee Museum. The protestors began at a professional meeting rather than through appeals to the public because they were confident they could persuade the Survey to change its policy. The Biological Survey was, after all, a scientific enterprise, with close ties to academic mammalogy. Specimens and scientists went back and forth and personal friendships reinforced professional ones. Neither side could afford, nor would want, a break. So the scientists reasoned, but their confidence was misplaced. The predator and rodent control program had changed the Survey's interests and leadership. It was becoming a service agency, and the woolgrowers, not the scientists, were calling the tune. The shift became evident even as the committee was working. In 1927 E. W. Nelson retired. His successor, Paul G. Redington, was not a mammalogist and not even from the Survey: he was from the Forest Service.[13]

For a number of reasons, the committee made no report until 1928. Heller was in Africa. Dixon was in the field, working on a survey of the mammals of Nevada that was to provide data for the committee. And a further delay was brought about by Nelson's retirement. The group decided to give Redington a chance to get settled in office.[14] But more fundamental problems remained. The most glaring was a lack of real evidence. The Survey claimed that coyotes were a menace to stock and that control operations helped the ranchers. Merriam had often spoken of the "toll" levied on the ranchers, and David Lantz, the Survey's expert on poison formulas, had said that "coyotes destroy considerable game, are widely known as a menace to stock," and that

the animal was "especially notorious as an enemy of the sheep industry." In 1907 Vernon Bailey said that each wolf and mountain lion cost
ranchers $1,000 a year, bears $500 apiece, and coyotes and bobcats
$50 per animal. W. B. Bell repeated these estimates fourteen years
later.[15]

Government agencies normally view their work with pride and the
situation facing them with alarm, but there are good reasons to believe
that the Survey's estimates were the range equivalent of old wives'
tales. No work had been done or was being done on the food habits of
large mammalian predators, in or out of the Survey, that would support these estimates. The ranchers, who were not inclined to minimize
their losses or to preserve predators, rarely paid more than $20
bounty for a wolf. This seems far too cheap for animals that were supposed to cause $1,000 worth of damage per individual per year.[16] In
private, Survey officers admitted that the figures had little basis in fact.
The 1928 PARC conference in Ogden, Utah, had a panel discussion
on "What Should be Our Uniform Method of Computing Damage
from Predatory Animals and Rodents and What is the Best Method of
Computing in Dollars and Cents the Savings of Control?" Only one of
those present thought it "comparatively easy" to assess damage.
Others spoke of the difficulties of gathering good data, the problems
of surveying stockmen, and the pitfalls of questionnaires. Ira N. Gabrielson bluntly stated that "[a]ny estimates of damages done or savings effected can only be the wildest guess work." They "will ultimately
do enough harm to the Bureau to more than overbalance any temporary notoriety and assistance they may occasion. . . . Such estimates as
must be made in order to satisfy Congress should be given as little
publicity as possible."[17]

The mammalogists, though, had little evidence to back their claims
that coyotes checked the growth of rodent populations. They took the
"common-sense" position that since coyotes ate mice (fact) they held
down the mouse population (assumption). It seemed plausible, but
that was all that could be said for it. This state of theory and knowledge is evident in the reaction of both sides to the great Kern County
mouse outbreak of 1927. In January 1927, a horde of mice that had
been breeding in the fallow fields of a dry lake bed in Kern County,
California, spread out across the country. They overran barns and
houses, and a slick layer of crushed rodents made roads impassable.
Stories circulated widely about housewives who had spent a week on

the furniture, never touching the floor. Traps and poison reaped an incredible harvest—at one warehouse authorities buried two tons of dead rodents.[18]

That was the situation, but what had caused it? E. Raymond Hall came to Kern County shortly after the mouse population had peaked and spent some time reconstructing the events. The abundance of food and shelter in the lake bed, he said, had been a major factor, but the lack of natural enemies was the key. In 1924–1925 the Survey had conducted a predator control campaign in Kern County, and farmers regularly killed coyotes, hawks, and owls. It was this slaughter of the animals that ate the mice, Hall said, that had caused the irruption. Stanley Piper, working for the Survey and the California Department of Agriculture, arrived a week after Hall, did his own investigation, and blamed the long grass in the lake bed. It had provided food and sheltered the mice from predators. It was unrealistic, he said, to blame humans or to rely on natural controls. Irruptions were natural phenomena. They had occurred long before predator control work or human settlement had come to disturb the "balance of nature." People had to manage the system—and he pointed to the Survey's work in poisoning the hordes.[19]

That two scientists could come to opposite conclusions from the same evidence suggests that something was awry. It was. The debate was not about science at all. Each side presented a picture of how humans were related to nature and how nature worked—a picture based on what the culture provided. Neither picture could be proved, both were simplified and possibly oversimplified. Piper presented a dichotomy: leave areas to the workings of nature or manage them. Man, he said, had destroyed the balance of nature. Hall argued that nature was normally in balance and that humans should rely on natural mechanisms of population control.[20]

No one could make the argument effectively because no one knew what regulated natural populations or how the mechanisms worked. Scientists were only beginning to work out research methods and concepts that would allow them to tackle this problem.[21] They were themselves living between two worlds. One was the old natural history, based on observation and qualitative studies and concerned with classification and distribution. The other was the new animal ecology, using quantitative and mathematical methods, investigating the relationships among organisms and their interaction. Grinnell, for example, had begun as a collector and had been director of the Museum of Vertebrate Zoology at Berkeley for six years before he received his Ph.D. degree in zoology. He was interested in subspecies and the distribution of characteristics and their correlation with the environment, but as a means of studying "evolution now in progress," not ecology.[22] He used ecological concepts and even made an important contribution to "niche" theory, but he trained his students in taxonomy and vertebrate zoology. He resisted ecological studies at the museum as well as ecological theories his students developed after they left.[23]

Of the committee appointed to investigate the predator control program in 1924, only one member, C. C. Adams, had a complete professional academic education, and his Ph.D. in ecology (though the degree said zoology) was one of the first given in this country. Goldman and Bailey were naturalists trained in the Survey. Heller had a B.A. degree from Stanford and had been on the staff of the Museum of Vertebrate Zoology, but had made his reputation on the Roosevelt African expedition of 1909–1910. Dixon had done some graduate work but, like Grinnell, only after he had established himself as a collector and naturalist.[24]

These circumstances, as well as professional differences, made it hard for the committee members to reach agreement on the predator control program. In fact, they presented two reports. One, signed by all, was a noncommittal recital of common ground. Predators, it said, had values—educational, scientific, and economic—and should be preserved. It suggested either the national parks or isolated parts of the public domain but did not recommend any specific areas. The second report, signed by Dixon, Adams, and Heller, charged that the Survey and the western livestock industry were conducting an extermination campaign against western wildlife that could not be defended on scientific or economic grounds. The estimates of stock lost to predators came from ranchers and state officials who had every reason to exaggerate the extent of the danger. The analyses of the stomach contents of dead predators was inaccurate and biased. Current policy, it said, should give way to a "system of intelligent controls" adapted to the particular needs of each part of the West.[25]

Redington replied to the second report in an open letter to Adams. He did not answer the charges, choosing instead to emphasize the "general agreement" between the committee and the Survey on the question of predators. They could not be allowed outside special sanctuaries; "civilization will require all space except those areas that are in advance specifically reserved for wildlife conservation." The Survey's operations, he said, would merely "hasten the inevitable." The supplementary report, he went on, criticized chiefly the lack of research done before control operations were started and the chance that the Survey was eliminating study material for the scientists. Research on control, though, was not needed. Predators clearly had to be curbed to protect stock and the work had to be done now. As for scientific data, the Survey conducted research on habits and distribution of predators while it carried on control operations.[26]

Redington did take pains to counter the charge that the Survey was "exterminating" wildlife. This was, even then, a delicate subject; no one openly advocated wiping out a species. Redington had already warned field agents that they should try to dispel the impression that predator control was doing away with wildlife. He called for public education and suggested that the word "extermination" not be used within the Survey.[27] He made the same point to Adams. The agency, however, had little success in getting rid of the word. Articles continued to refer to "wiping out" predators and "getting rid of" the "stock killers." Redington himself had talked of "extermination" when testifying before Congress in 1927, and in 1931 the head of the House subcommittee dealing with agricultural appropriations was clearly under the impression that predator control was designed to "get rid of" the varmints.[28] The most embarrassing incident came in 1929, when Jenks Cameron's *Bureau of Biological Survey* described the predator control campaign under the name "extermination." Goldman, in a discussion with Adams, denied the charge. The author, he said, had no position with the Survey and was probably a "hack writer" (a species even the editor of the *Journal of Mammalogy* thought might be wiped out without loss to the community).[29]

At this point another of Grinnell's friends, A. Brazier Howell, entered the dispute. Complaining that the mammalogists had been handling the Survey with "kid gloves," he wrote a petition calling for an end to the predator control program as currently constituted. He circulated it among scientific institutions and societies and sent it to Redington in April 1930, with a covering letter and the signatures of 148 scientists. The petition and the covering letter were not representative of the full scientific community. The petition went to only a few institutions; not all who signed were mammalogists (Alfred C. Kinsey certainly was not); and the covering letter was Howell's alone. Still, the petition and the letter are worth some analysis. They had arguments Howell thought worth making and that his colleagues supported, if only in outline.

The Survey's work, the petition said, was an imminent danger to the "very existence of all carnivorous mammals, including those valuable species which constitute the chief check upon injurious rodents and are a vital element of our fauna." These strong words did not mean that the scientists were against predator control or that they wanted to preserve the larger predators throughout the country. We do not, Howell said, "deny that control of predatory mammals is advisable in certain instances and in certain places; only that it is greatly and dangerously overdone. Also we make no mention of wolves and mountain lions which, whatever their values from an aesthetic viewpoint, are truly killers and are destructive. Our claims are based on the economic viewpoint alone."[30]

Howell challenged all of the Survey's claims for the program. He pointed out that before the agency had started predator and rodent poisoning, it had appealed to the balance of nature, arguing that coyotes kept rodents in check. It had neglected to do so in recent years. The Survey claimed that analysis of the stomach contents of predators proved coyotes were a serious problem, but the head of its own food-habits laboratory, W. L. McAtee, refused to accept government trappers' reports as valid scientific evidence.[31] The bureau said that poisoning and trapping took a negligible toll on nontarget species. Field agents, Howell pointed out, had no incentive to look for these animals and every reason to ignore them. Joseph Dixon reported that there were large losses of furbearers, and Dixon, Howell said, was the best observer the Survey had in California (Dixon, like many of the mammalogists, including Howell, had worked for the Survey). Whatever the intentions of officials in Washington, Howell charged, the men in the field were exterminating native wildlife.[32]

Besides mobilizing the scientists, Howell and his friends appealed to the public. Howell published two articles in *Outdoor Life*, "The Borgias of 1930," and "The Poison Brigade of the Biological Survey." W. C. Henderson replied in "The Other Side of the Poison Case." Hall weighed in with a two-part article, "The Poisoner Again." Harold Anthony of the American Museum of Natural History blasted the Survey in *Science*. Howell reinforced it, and then Goldman replied.[33]

The mammalogists, though, were swimming upstream. Even as Howell presented his petition, the Survey was asking Congress for $1 million a year for the next ten years to wage an all-out campaign against predators. This, it promised, would reduce predator populations to a very low level, and Congress could then reduce appropriations. Late in April 1930 the House and Senate Agriculture Committees heard testimony on the first part of this plan, a bill to give the Secretary of Agriculture authority to conduct work for the "eradication, suppression, or bringing under control" of injurious animals on public and private land. Western congressmen and senators opened the hearings with vivid descriptions of the hardships their constituents were suffering. Ranchers and representatives of the National Wool Growers Association added their support. The Survey produced an impressive array of facts and figures.

Howell and Hall testified against the program and primed friendly congressmen with questions for the Survey's witnesses, but they could do little. Even though there were deviations from approved field practices, isolated protests against poison, and deaths of some nontarget animals, Howell and Hall could not show that the program was economically unsound or biologically dangerous. Since they were not calling for an end to predator control or proposing an alternative plan, the Survey was able to meet many of the attacks by promising to correct whatever problems did exist. The agency's greatest advantage, though, was that there was little interest in preserving "useless" animals and a strong and effective lobby pushing predator control.[34]

The Survey did not want the mammalogists as adversaries—too much of its work depended on their cooperation—and Goldman and Henderson came to the American Society of Mammalogists' meeting the next month to explain and defend the program. Goldman, in a speech entitled "The Coyote—Archpredator," painted a bleak picture of western ranchers beset by a "bold and ruthless marauder." Henderson used stomach-content studies to show that coyotes were fond of beef and mutton. They met a barrage of criticism. Hall claimed that Henderson's work was full of errors, from the food studies and the trappers' reports to the work on the effects of large doses of poison on small animals. Dixon and Hall said that the stomach-content studies had a systematic bias. Much of the evidence came from animals killed by eating poisoned bait. And in any event, the government hunters were untrained and careless; their data was useless. Hall and

Adams said that the government had refused to cooperate with out-side scientists and would not make a clear statement of policy. Howell charged that the Survey was in the woolgrowers' pocket. The commit-tee appointed in 1929 to review the 1928 reports said that the current program could not be scientifically justified. It issued another call for change.[35]

The mammalogists and the Survey did agree on a joint inspection team to look into allegations that field men were routinely violating guidelines. Mutual suspicion put an end to that effort. Goldman told Howell that he knew what he would find. Howell suspected a cover-up. Rumors circulated among the field agents that the inspection team might fire people (a serious concern during the Depression). Some supervisors were openly hostile to the inspectors. The trip raised more questions and tempers than it settled. The mammalogists repeated their earlier charges that the Survey was conducting an indiscriminate campaign against native wildlife and "drumming up" business. The Survey stuck to its guns.[36]

The next year the use of thallium-poisoned grain for rodent control added fuel to the fire.[37] Thallium was a cumulative poison, unselective and slow-acting (poisoned animals might take days to die), but the Sur-vey and the California Department of Agriculture decided to use it to control ground squirrel populations now "bait-shy" from sublethal doses of strychnine.[38] In May 1931 Jean M. Linsdale, one of Grinnell's students and co-workers, published an article in *Condor*, "Facts Con-cerning the Use of Thallium in California to Poison Rodents—Its De-structiveness to Game Birds, Song Birds, and other Valuable Wild Life." The facts, Linsdale said, were that thallium-poisoned grain put out for ground squirrels was killing millions of other animals. In an accompanying editorial Grinnell said he had read the article with "a peculiar feeling of despair." Over a third of the state was being poi-soned for the benefit of a small group of people. As many as fifty mil-lion animals and birds may have been killed. The Survey was moving "toward a 'practical' type of 'conservation' " that was destroying wild-life.[39]

The PARC fought back. Stanley Paul Young, in charge of the West Coast office, sent out a team to check conditions in the field. He met with representatives of the Western Society of Naturalists and the Au-dubon Society.[40] He pointed to the agency's research work, the close supervision it exercised over field operations, and its cooperation with the California Department of Agriculture. The PARC also took the offensive, supplying information for a California Department of Agriculture bulletin that took Grinnell and Linsdale to task, showing, with quotations, that both men had previously supported ground-squirrel control and bemoaned the losses the rodents caused to California agriculture. Why did they now object? It denied that the poisoning caused enormous losses among nontarget species; Survey agents following the poisoning crews had not discovered the carnage that the biologists had reported.[41] Grinnell and Linsdale had apparently gone too far. When they introduced a resolution to condemn thallium poisoning at a meeting of the Cooper Ornithological Club, it was voted down, and Dixon and Tracy Storer, a former Grinnell student, were among the majority.[42]

The Survey's victory was part of a general collapse of the opposition. Howell and Hall had blocked the Survey's plans in 1930, but the next year Congress passed the Animal Damage Control Act.[43] It gave the Secretary of Agriculture increased authority to carry on control operations, approved the ten-year plan for all-out control (though Congress never funded it at the levels envisioned), and provided the PARC with a statutory authority that remained its legal charter into the 1970s. The Depression also helped the program. As profit margins shrank, farmers and ranchers were even more anxious to reduce their losses and eager for government help. Spreading poison grain got rid of rodents and also provided relief work in rural areas. The scientists hoped that the Roosevelt administration would be more sympathetic. It was not, and the dissidents were reduced to writing letters protesting the use of Civilian Conservation Corps boys to spread poisoned grain.[44]

There was even a reconciliation of sorts. In 1931 Ira Gabrielson resigned from the American Society of Mammalogists to protest the Society's condemnation of the Survey's predator control work. Four years later, when he was appointed chief of the Bureau of Biological Survey, he wrote to request that his membership be reinstated. Harold Anthony, then secretary, sent the letter to Howell. Gabrielson, he pointed out, was not offering to pay his back dues. He concluded, though, that it would be best to let him back in without a fuss, and that was done.[45]

The PARC's triumph was complete, but it was hollow. The agency had fought off the mammalogists, but it had not convinced even those within its own ranks that the program was sound. In January 1931 Olaus J. Murie, a PARC supervisor in Jackson, Wyoming, wrote a long letter to W. C. Henderson, assistant chief of the Survey. He wanted, he wrote, to "get things off my chest." Murie was a field biologist who had started with the Survey in 1920; he should have been one of the committed. Instead, he was troubled by the Survey's handling of the situation. On the trip with the joint inspection team he had found people trying to convince him that the Survey was right, rather than giving him the information he needed to make a decision. They had even offered him bribes—"the chance to hunt ducks without a license, etc." He had reviewed the mammalogists' complaints and the Survey's response, and "to tell the truth I could not see that the queries of the Society had been adequately met."

His complaints echoed those of the dissidents. The agency, he pointed out, had started to kill predators in Alaska without benefit of scientific studies. It had relied on the "usual run of sourdough information," which was being "accepted at face value." Wolverines were being trapped, but "I have yet to meet anyone who knows anything about wolverine food habits in a quantitative way." Are we, he asked, increasing the list of predators "just because we are on the ground and in the business?" He knew these views were not popular, for he concluded by asking, "Am I a black sheep in the Bureau fold now?"[46]

Five months later he wrote to A. Brazier Howell. It was, he stressed, a private letter; Howell could show it to Harold Anthony (Murie's friend from college) and to E. A. Preble, one of Merriam's old crew now nearing retirement—but only to them. He was, he confessed, "very fond of native mammals, amounting almost to a passion," and "would make considerable sacrifice for the joy of animal companionship and to insure that other generations might have the same enjoyment and the same opportunity to study life through the medium of the lower animals." He thought the cougar was "Nature's masterpiece in physical fitness," the wolf a "noble animal, with admirable cunning and strength." He wanted even the "so called injurious rodents around." He believed in control, he said, but less than was now being used, and he wanted the work put on a sound basis. We are "passing around an appalling amount of misinformation about the effects of predators on game. I have been awakening to this fact only in the last few years."[47]

He said that he hoped in the next few years to work out his ideas.

He did. In 1945 he left the Survey, which had by then become the Fish and Wildlife Service, to work for the Wilderness Society. Seven years after that, in a letter to a friend who had remained in the PARC, he reflected on the events that were now some twenty years in the past. The "scientists who became so concerned at that time did not, I believe, understand their own motivation. . . . The big issue put forth was that 'innocent animals' were being killed incidental to poisoning operations. Deep in their hearts, if they had thought it out fully in those formative years of the opposition, was concern for the coyote itself. . . ."[48]

But how could they have "thought it out fully"? They, almost as much as the rest of the population, brought to the questions of predation and predators a load of cultural baggage and almost no analytical tools. They loved nature and animals, but, like Seton, saw no alternative but destruction. It "was inevitable that the wolves should be hunted down. They were entering their twilight in a country being converted from wilderness to pastureland."[49] Why should they be saved, aside from Romantic feelings about the "joy of animal companionship?" We now argue that wolves and coyotes need to be saved because they are parts of a system. All are essential and all must be saved together. Even Romantic feelings about the wolf as a symbol of the wilderness rely on its position as a controlling element in the ecosystem. Grinnell and his friends did not have the science that buttresses that view. It would only develop in the next decade, as animal ecology found research tools and theories to suit its subject, and as game management began to form a wildlife policy based on them.

PART II

A Time of Transition

The mammalogists' battle with the Division of Predator and Rodent Control showed that some people's ideas about predators and their place in America had already changed. Other people would follow suit, as a set of changes in the interwar period made the value of a varmint more obvious. Animal ecology developed as an independent discipline in the 1920s. Then and in the next decade it spread its ideas, first to people in the related field of game management, and then to the general public. Books and magazines and, after World War II, movies and television carried the message to a new generation. By 1960 all that remained was to translate the new understanding and concern into law and policy.

CHAPTER FIVE

Making a New Wildlife Policy

1920-1940

So far we have the scientist, but not his science, employed as an instrument of game conservation.—Aldo Leopold, *Game Management* (1933)

By the 1920s the program that hunters and wildlife conservationists had urged on legislatures had been in place for twenty to thirty years. The results were disappointing. States had cut bag limits, enforced game laws, encouraged the destruction of predators, and introduced or reintroduced game species, but the hunting was no better than it had been. In some cases it was worse. Ducks had been subject to federal regulation since 1918, but the 1931 duck season had to be cut from three and one-half months to one month to preserve the few birds that remained. Bobwhite quail, the most popular upland game bird, were so scarce that some state legislatures reclassified the species as a songbird and ended its hunting altogether. Deer populations rose and fell in strange ways. Starvation and disease struck seemingly secure populations. There was policy—in the sense of agreed-upon actions—but it was not working.

The most spectacular failures were in deer "management," and among these the fate of the herd in the Kaibab National Forest on the North Rim of the Grand Canyon stood out. In 1906 the federal government made it a game preserve. The herd grew rapidly until the mid-1920s, and the policy of protection seemed a success. Then the animals began dying and the herd continued to decline for a decade. "The Kaibab" became a classic conservation horror story, repeated in sportsmen's magazines and game management and ecology texts for the next forty years. More than any other single incident, it forced a reevaluation of management methods and the theory on which they were based. And, though there were no wolves and few other predators in the forest, the episode helped shape ideas about predation and predators.[1]

The Kaibab plateau, some 1,200 square miles on the North Rim of the Grand Canyon, was part of the area set aside as the Grand Canyon Forest Reserve in 1893. It was the bulk of what became in 1906 a game reserve and in 1908 the Kaibab National Forest. Alarm about declining game was at its height, and the policy was to give the deer as much protection as possible. Ending hunting was only the first step. The Forest Service also cut the number of cattle and sheep grazing on the plateau and called in the Biological Survey to "control" predators. It did. Between 1906 and 1931 the Survey's hunters "removed" 781 mountain lions, 30 wolves (all that remained), 4,849 coyotes, and 554 bobcats.[2]

The policy seemed to work; each year there were more deer. In 1920 the forest supervisor reported a new problem: too many deer. Forest Service biologists confirmed it. There were so many animals that they were eating all the new growth. Like the mice in Kern County, they were eating themselves out of house and home. Unlike the mice, they could not spread out over the countryside; there was the Grand Canyon on the south and deserts on other sides. Two years later another inspection team, this one accompanied by the Survey's E. A. Goldman, recommended reducing the herd to avoid damage to the range. The next year E. W. Nelson, chief of the Survey, made the same plea. In the fall of 1923 and the summer of 1924 biologists again reported that the deer were causing serious damage; Goldman said that conditions were much worse than they had been the year before.[3]

All the evidence was impressionistic; the Forest Service did not even know how many deer there were. It had begun taking censuses of game in 1914, but with more zeal than accuracy. A ranger later described the process as educated guessing. Each ranger, he said, figured out the game population in his area by combining what he knew with what other people told him. The next year, because things were getting better under his management, he increased the number. Growth continued until fire, flood, disease, or a bad winter allowed him to drop back and begin building his herd again.[4] Estimates, then, might vary widely. For the Kaibab herd at its peak they ranged from 20,000 to 100,000.[5]

Whatever the numbers, they were too many. Every scientist who visited the forest, then and later, recommended cutting the herd to preserve the range. The obvious way to cut the herd was to kill deer, but there were serious obstacles to this. For one thing, the Forest Service, though it administered the area, had no control over the deer. Migrating from summer to winter range and back again, they spent part of the year in the national forest, part in Grand Canyon National Park, and all of it in the state of Arizona.[6] Governor E. P. Hunt and state game officials agreed that hunting was needed, but Steven Mather, director of the National Park Service, did not. There was, he thought, no shortage of forage and no excess deer. He vehemently opposed any killing. The herds were a tourist attraction that should be preserved.[7]

The Service had to take Mather seriously. He was a master of public relations, taking a popular stand. For two generations, sportsmen's magazines and leading hunters had warned that there were too few deer and that kills had to be reduced. They had preached game preserves, predator control, and the "buck law."[8] To reverse that, to accept, even sponsor, a large deer kill—and nothing else would do—was a daunting prospect. C. E. Ratchford, the Forest Service's Inspector of Grazing, wrote to Chief Forester William B. Greeley that "we have here a very big problem. . . . Severe criticism of the Service by certain game enthusiasts will result unless the problem is handled with all the tact and ingenuity we are capable of putting into it."[9]

The Forest Service had arranged for a hunt on the Kaibab in the fall of 1923, but Mather's opposition caused a postponement. The foresters sent the problem up to Secretary of Agriculture Henry C. Wallace (the Forest Service was in his agency). Wallace resorted to a common bureaucratic maneuver. He appointed an investigating committee, composed of representatives of all major users of the area.[10] In the summer of 1924 committee members toured the area. They were shocked; conditions were the worst any of them "had ever seen."[11] The forest looked as if it had been mowed, in Aldo Leopold's later description, "as if someone had given God a new pruning shears and forbidden Him all other exercise."[12] As high as a deer standing on its hind legs could browse, every twig, branch, and green growing thing was gone.

The group recommended that half the herd be removed. The Forest Service promptly set to work, but publicity, some of it generated by Mather, had caused Governor Hunt to change his mind. He decided that there were too few deer, and he championed the cause of his angry constituents.[13] When the Forest Service issued permits, Arizona officials challenged the government's right to set hunting regulations. The Forest Service then tried to reduce the herd without killing deer. It contracted with a rancher who promised to herd and drive the animals off the plateau. They rounded up eighteen deer, a feat that earned a headline in the *Tucson Star*: "Huh, White Man Heap D__ Fool, Says Navaho Chief to Game Warden, Who Brands Attempt 'Complete Fizzle.' "[14]

In the winter of 1924–1925 cold weather and the lack of food accomplished what the Forest Service could not. In the spring, deer carcasses were scattered throughout the forest. That summer Goldman and E. Raymond Hall found stunted animals whose bones could be counted through the skin. The Forest Service concluded that the only solution was to kill the deer, preferably by hunting. It asked the state to suspend its regulations on the Kaibab and allow hunters to kill more than one deer apiece. Politics kept officials from agreeing; the need to cooperate with the Forest Service on many other issues kept them from fighting too hard. They arranged a test case. Secretary of Agriculture Jardine issued new regulations allowing the Forest Service to manage deer on its land. The Forest Service began killing the Kaibab deer under new regulations. Arizona attempted to stop it. The United States sued to enjoin the enforcement of state game laws on the Kaibab.[15]

When *Hunt v. United States* reached the Supreme Court in 1928 the justices, unexpectedly, ruled against Arizona. In a decision that opened the way for federal control of game, and ultimately all wildlife, on federal land, they declared that the "power of the United States to thus protect its lands and property does not admit of doubt, the game laws or any other statute . . . of the state notwithstanding."[16]

The *Hunt* decision gave the Forest Service power to trim the herd, but it did not make it any more popular among Arizona sportsmen. In 1927 Will C. Barnes, Assistant Forester, had reported to Ratchford that he had been "in Arizona last summer [and] found everyone was against us on the killing. All my old friends down there felt we were determined to destroy this herd. . . . Everybody in that part of the country looks upon the buffalo and deer herds as two of their greatest natural possessions, and they are determined that as far as possible they shall be kept intact."[17] After the *Hunt* decision the Forest Service raised the bag limit. This made it even less popular. In October 1930 the Phoenix *Gazette* published an editorial, "That Kaibab Myth," attacking the Forest Service. The "myth," for the editorial writer, was that the range was overcrowded. There were, he said, too few deer, and these liberal limits were allowing "a steady stream of game hogs" to destroy "the greatest deer forest in the world . . . taking away from future generations the same privilege of fine hunting that we have today."[18]

By then other areas were going through the same cycle of rapid growth and abrupt decline. Irruptions had been observed before, particularly among lemmings and mice. They had been particularly marked in cases where (as the rabbit in Australia) a rapidly breeding animal had been introduced into an area where it had few natural enemies. Deer irruptions in the United States had not been unknown, but they became more common, frequent, and severe after 1920. Aldo Leopold and co-workers found two irruptions in North America between 1880 and 1899, two more in the next decade, four between 1910 and 1919, then a quickly rising curve. There were three in the next five years, four in the five years after that, seven in the first half of the 1930s, seven more in the second half, and fifteen between 1940 and 1945. All three deer species (white-tailed, black-tailed, or mule deer, and Columbian) and every part of the country but the Southeast had its share (Leopold and co-workers thought that screw worms and hound dogs combined to check the southern herds).[19]

Remedial action, the same survey reported, almost invariably came too late and did too little. This was not due to a lack of scientific advice. States and federal agencies regularly called on Survey biologists and outside experts. The team that reported on the Kaibab in 1931, for example, included Dixon and Hall from the Museum of Vertebrate Zoology in Berkeley; Homer Shantz, a Bessey student in plant ecology who was then president of the University of Arizona; Goldman from the Survey; and Ben Thompson from the National Park Service.[20] The scientists, though, could do little more than state the obvious: there were too many deer. If they were not removed they would cause serious long-term damage to the range. The quickest way to remove them was to have hunters kill them.

No one proposed a return to natural predation as a check. In a "Personal and Confidential" letter to George Bird Grinnell, E. W. Nelson wrote that the Kaibab was a disaster area. Overbrowsing was killing almost all new growth. Were we going to have, he asked, conservation by "common sense or, . . . sentimental ignorance which refuses to recognize facts and which persists [sic] that nature take its course." That deer had once been abundant had no bearing on the current situation. Then "predator animals roamed at will, mountain lions were exceedingly abundant and the Kaibab Plateau was one of the principal hunting grounds of the Navaho Indians. . . . At the present time the mountain lions have been almost completely exterminated in this area. No hunting is permitted by the Indians or others. . . ."[21]

Here, however, he drew back. He did not go on to propose what seems to us the obvious conclusion: that predators be brought back to the area. Nor did anyone else. The 1931 committee was the only group even to raise the subject; its report gingerly referred to a "theory" that if predators were not trapped they would prune the herd back to the forest's capacity to support it. The common assumption was that predators would not check a herd but would destroy it. In his 1929 survey of the mammals of the Grand Canyon (an area that included the Kaibab) Vernon Bailey wrote that it was important to keep predators down, and no one challenged him.[22] Direct human management was the only acceptable—indeed the only "thinkable"—course. Certainly it was for the Forest Service, still issuing permits to trap coyotes, bobcats, and mountain lions, and for the Survey, which continued predator control on the plateau until 1931.[23]

New Foundations for Managing Game

The failure of "cut and try" to produce game led people to look not just for better policies but for a better basis on which to build them. That turned out to be animal ecology. Even as deer herds surged out of control and collapsed, animal ecologists were beginning to make headway in understanding these problems. They were discarding models borrowed from plant ecology and finding research techniques and concepts suited to their subject. Charles Elton, an English ecologist whose writings affected both animal ecology and, through Aldo Leopold, game management, said of this period that he and his colleagues were just beginning to enter "a new mental world of populations, inter-relations, movements, and communities," to study "living nature in the field" and to reject "easy generalizations about adaptation and the balance of life."[24] The new approach was not a magic wand or a complete answer. Indeed, even today wildlife biologists often hesitate to prescribe policies and acknowledge that they know too little to be certain about a particular situation. Still, the theories and research methods of the 1930s were a major step forward, and they would help make game management a profession and an applied science. They would also stimulate thinking about predators and pre-dation. Work done by managers fed back into ecology and, because game managers depended on public support, they also diffused the ideas of ecology and their management strategies based on ecology to the public.[25]

Elton's *Animal Ecology* offered game managers two important pre-scriptions. One was a set of concepts that directed research, and di-rected it into new channels. The other was quantification and field methods that encouraged quantitative studies. The concepts—food chains, the food cycle, trophic levels, and niches—put flesh on the skeleton of the "balance of nature." They provided ways to analyze animal communities and their relation to plants, to order data, and to classify communities. They provided, in short, theories that could be used in the field and which could guide policy. Quantification intro-duced new elements of rigor and forced attention to fundamental def-initions and limits.

New concepts led to new questions. One effect of the new ideas was to focus attention on predation and predators. Much of the analysis of terrestrial animal communities was couched in terms of food chains. These were, in the last analysis, studies of who ate whom, and wolves and coyotes were conspicuous consumers. They were also at the end of the food chains in their particular ecosystems, a circumstance that made them even more interesting to scientists working out the dynam-ics of populations and the problems of building deer herds for shoot-ing or appreciation. The new ideas also shifted scientists' attention from a concern with the relations between species to studies of partic-ular populations in given environments (a subtle but significant change). Instead of impressions they relied on censuses. Rather than the composition of the predator's diet, they considered the effect of predation on the prey population.

"I do not know," Aldo Leopold wrote, "who first used science crea-tively as a tool to produce wild game crops in America. . . . The idea was doubtless conceived by someone long before it was first success-fully applied by the Biological Survey in Georgia."[26] No doubt it was, but it was the Survey's project, and the book that came from it, Her-bert Stoddard's *The Bobwhite Quail: Its Habits, Preservation, and Increase*, set the example and stimulated a more rigorous science of game man-agement.[27] The study began as a very practical piece of work. In the spring of 1923 a group of wealthy New York sportsmen asked E. W.

Nelson for help. Game on their Georgia quail plantations had been declining for the last few years. Could the Survey find out why and tell them what to do? It could, and they soon agreed on a cooperative project. The sportsmen donated a shooting plantation near Thomasville, Georgia, for the experiments and the funds for a three-year study. The Survey promised to find a man to do the work. Results would be made available to the owners and, when published, to all sportsmen.

Nelson chose Herbert L. Stoddard to do the work. That seemed to reinforce the project's practical orientation. Certainly the Survey could not be accused of choosing academic knowledge over field experience. Stoddard had dropped out of school at fifteen to become a farmhand. He had turned his youthful interest in natural history into a profession by becoming a collector and taxidermist, working first for the Milwaukee Museum (under the direction of Edmund Heller) and then for the Field Museum in Chicago.[28] He had no formal training in ecological theory and never developed much taste for it. When, in 1931, Leopold lent him a copy of Elton's *Animal Ecology*, Stoddard read it and told Leopold that "I must make a point of getting his books; his comes nearest of being the sort of ecology I can appreciate."[29]

Still, Stoddard used methods and asked questions that Elton and other ecologists would have recognized and approved. He counted coveys, quail nests, and clutch sizes. He calculated percentage of nesting success, extent of renesting, and the age composition of the population. He assessed mortality among eggs, chicks, and adults. He calculated the effect each predator species had on population growth. It was the whole apparatus of quantitative ecology. He also banded birds to trace ranges and mobility, mapped covey territories against different kinds of cover and food sources, and burned areas to study the effects of fire and regrowth. *The Bobwhite Quail* is full of practical suggestions for raising game, but they are grounded in observation and calculation.

Others soon followed, using other species and finding substitutes for rich plantation owners. States anxious to keep or increase money from visiting hunters and tourists began to pay for expert advice. The Sporting Arms and Ammunition Manufacturers Institute, which had an obvious stake in sport hunting, began funding projects. The Survey, as in Georgia, helped direct the studies. In 1934 W. L. McAtee reported that the Survey's Division of Food Habits Research was co-operating or supervising work on Hungarian partridge in Michigan, bobwhite in Wisconsin, ruffed grouse in Minnesota and New York, scaled quail in New Mexico, and Gambel's quail in Arizona.[30]

Research concentrated on game birds because of their economic importance, but the work had larger implications. Here one study stands out: Paul Errington's on the population dynamics of the northern bobwhite quail. Not only was it the first extensive study to show a situation where predation was not a major check on a game species, but it launched Errington on a course of research that would make him his generation's expert on predators and predation in mammalian and bird populations.

In 1929 Aldo Leopold, then on a temporary appointment teaching game management at the University of Wisconsin, decided to duplicate Stoddard's work at the other end of the bobwhite's range, in south-central Wisconsin. He assigned the task to Errington for his Ph.D. research. Lacking a plantation and sportsmen, Leopold chose five square miles of farmland near Prairie du Sac, Wisconsin, twenty miles northwest of Madison, and got funds from the university and ammunition manufacturers.[31]

Errington began with basic quantitative data. Tramping through the fields, pastures, and woodlots, he counted every covey each week from the time it formed in the winter until it dispersed in the spring. He took notes on food, cover, snow and snow crust, ice, wind, temperature, quail mortality, and the causes of death. The data from two, then three, years of study showed an unusual pattern. Survival rates seemed much more closely linked to the area where the covey wintered than to the predators that preyed on it. An influx of gray foxes in the winter of 1933–1934 tested and confirmed this. Land that had carried many quail through the winter did so again, despite the foxes, and land that had not, did not. "Within ordinary limits," he cautiously concluded, "the kinds and numbers of native flesh-eaters may not be of much consequence in the winter survival of wild northern bobwhite populations."[32]

Errington left Madison in 1933 to take a job at Iowa State University, but he continued to work on predation and its role in population regulation. People's ideas, he came to realize, were based on extrapolation. From the known *fact* of predation they leaped to the unknown *effect* of predation, from the knowledge that some animals ate others to the assumption that this controlled the population.[33] This did not have to be the case. At Prairie du Sac, for example, it had not been; food, cover, and occasional severe winter storms had been the limiting factors. The birds had not lived in terror of their enemies, nor had they bred frantically to keep ahead of relentless thinning by the flesh-eaters. Errington did not believe he had found a single factor that controlled population growth. In fact, he pointed out in a paper he wrote with Stoddard that the importance of predation varied among quail populations. It may have played a major role among Stoddard's populations, but it did not among his.[34]

The research was most useful for the destruction it caused. It cut the ground from under the standard notions about predators and prey. Leopold was one of the first to call attention to this. In *Game Management* (1933), he relied heavily on Errington's work to argue against predator control as a standard practice. It might be necessary, he said, but it had to be shown to be correct in each case. It could not be assumed that killing predators would do any good. Like Errington, he did not think there was a simple answer. In our current state of ignorance "[t]here is only one completely futile attitude on predators: that the issue is merely one of courage to protect one's own interests, and that all doubters and protestants are merely chicken-hearted."[35]

Quantitative study of communities changed more than the way scientists looked at game birds. It marked a new phase in research on larger animals, including wolves and coyotes. Compare, for example, the work done by the Murie brothers, Olaus and Adolph, in the 1930s. Olaus, the elder by ten years, studied the coyotes of Jackson Hole, Wyoming, for the Biological Survey between 1927 and 1932. Adolph, working for the Park Service, looked at the coyotes in Yellowstone National Park between 1937 and 1939 and the wolves of Mt. McKinley from 1939 to 1941.[36] Their coyote research was about as close as two such projects could be. The populations overlapped and the areas were similar. Both studies began in response to complaints that predators were killing off the more desirable species. They had the same purpose—to find out what effect coyote predation had on elk populations under wilderness conditions. Even the conclusions were the same: individual elk were vulnerable, but coyotes were no danger to the elk population. Neither agency, to continue the similarities, liked the results. Olaus's colleagues in the PARC division of the Survey thought he had not sufficiently emphasized coyotes' appetite for mutton. Park Service officers who thought that "innocent" animals needed protection wanted to fire Adolph.[37]

Their research techniques, though, were very different. Olaus, in *Food Habits of the Coyotes of Jackson Hole, Wyoming*, was concerned with the coyote's diet. He analyzed stomach contents and scats and reported his results in terms of what the coyote ate. Adolph asked directly what effect predation by coyotes had on the prey population and studied a broad range of environmental factors to arrive at his answer. He looked at weather, food, disease, cover, and historical patterns of wildlife abundance. He counted elk and checked calf survival. He looked at coyote behavior and coyotes' relationships with prey and nonprey species, which ranged from bison and moose to porcupines and squirrels. His work was as comprehensive as the title, *Ecology of the Coyote in the Yellowstone*.

The change in methods is even more apparent in his study of the wolves and Dall sheep of Mt. McKinley National Park.[38] To assess the wolves' impact on the sheep, he reconstructed historical patterns of abundance of wolves, sheep, and buffer species (alternative sources of food). To test the argument that predators selected old, weak, ill, or otherwise defective individuals, he collected over eight hundred sheep skulls and compared the living and dead populations in terms of age, sex, known disease, and injuries. He went well beyond the problem originally assigned—whether the wolves were a danger to the Dall sheep—to speculate on the impact of wolf predation on sheep evolution.

The Wolves of Mt. McKinley was a pioneer effort in the ethology as well as the ecology of wolves. Murie spent much of the summers of 1940 and 1941 watching wolves at their dens.[39] What he saw surprised him. "So far as I am aware," he wrote, "it has been taken for granted that a wolf family consists of a pair of adults and the pups." One den, though, had one extra adult and another had three. In one case the wolves spent two years with the denning pair, which "suggests that it may not be uncommon." Close observation did not support the legends of savage, morose, "lone" wolves. "The strongest impression remaining with me after watching the wolves on numerous occasions was their friendliness." There was an "innate good feeling." They also had a complex social life and were markedly different in markings and behavior. With practice any one could be picked out from the group. Adolph included in his report sketches (done by Olaus) of "The Dandy," "Grandpa," and others.[40]

The brothers' research was not different because they had different ideas about wildlife and wildlife preservation; it was different because of their training. Olaus, born in 1889, had graduated from college in 1912 and worked on field and museum jobs for a few years before joining the Survey. His career pattern was much like the older generation of naturalists. Adolph, ten years younger, was one of the new academically trained specialists. He received his B.A. in 1923, and though he did some natural-history work—including a faunal survey in Alaska with his brother in 1922–1923—he went on to graduate school. The place he chose, the University of Michigan, and the man he worked under, Lee Dice, were in the forefront of ecological studies and wildlife preservation. Michigan was one of the institutions contributing to Howell's petition in 1931 against the Survey's predator poisoning. Adolph Murie, Lee Dice, university president Alexander Ruthven, and twenty-four other faculty and staff signed it. After getting his Ph.D., Adolph stayed on for five years at the zoology museum, working with Dice and others on various biological problems, including the irruption of the moose population on Isle Royale, an island in Lake Superior. Olaus got a Master of Science degree from Michigan in 1927, a year before his brother, but his graduate work was an interlude in his career in the Biological Survey.[41]

Ecology quickly spilled over into game management. Wildlife people seeking reasons for the failure of the older methods and those looking for a scientific base embraced the new science. Game management developed as an academic discipline in the 1930s, building its own degree programs, societies, and journals. It was an ideal time. The old methods had failed; there seemed to be a scientific base for action; and the public was interested in game. The mantle of science gave game managers legitimacy.

Game management developed within the land-grant university system and on the model provided by earlier sciences. This was a conscious decision, a deliberate application of an existing model. Aldo Leopold, for example, defined game management as "the art of making land produce sustained annual crops of wild game for recreational use." He thought its "nature . . . best understood by comparing it with other land-cropping arts."[42] He carried over the conservation ideals that had dominated the forestry program in which he had been trained—maximum production, efficiency, and expert management.[43] He also carried over the commitment to public education that the model required. Applied sciences had flourished in the land-grant schools by getting public support, and they had heavily relied on education, particularly extension education, to get their message across. Leopold, again, was quite conscious of this; in his plans for his position as professor of game management at the University of Wisconsin, he suggested using demonstrations, farmers' short courses, lectures, and exhibits to educate farmers, sportsmen, and nature lovers.[44] He was not alone. In 1942, discussing the "Viewpoint of Employers in the Field of Wildlife Conservation," William Van Dersal pointed out that these people wanted graduates who could write and speak good English. They had to meet and convince the general public. Tact and appearance, he warned, were as necessary as technical knowledge.[45]

Educating the public meant preaching from the text of ecological research. Game managers pointed out to farmers that weeds and underbrush sheltered and fed quail, pheasants, and rabbits, that the pothole in one field sheltered ducks, the patch of swamp in another was home for muskrats and mink. They told hunters of the futility of trying to increase game birds by shooting hawks and crows. On predators and predator control, the Kaibab came to the fore, as a (literal) textbook example.[46] Animal ecology books usually had a graph of the Kaibab deer population over time. It began at a low level early in the century, rose steeply to a peak around 1924, and declined even more swiftly after that. *Principles of Animal Ecology* labeled the curve "The effect of removal of predators on populations of deer on the Kaibab plateau in Arizona."[47] The most popular postwar textbook in game management, Trippensee's *Wildlife Management: Upland Game and General Principles*, had this to say:

PREDATOR CONTROL IN WILDLIFE MANAGEMENT. *Reduction in predaceous animals is one of the means by which wildlife managers have attempted to increase different species of game. Sometimes it has been successful, but sometimes it has given results that were more detrimental than the damage caused by the predator. A case in point is the Kaibab deer range after too effective control of mountain lions. In this instance the deer herd increased to the point where the range was severely damaged.*[48]

Magazine articles and popular books preached the same message. In *Our Wildlife Legacy* (1954) Purdue biologist Durward Allen (like Adolph Murie a product of the University of Michigan) said that the Kaibab was "only the type case; the same thing happened in many

77

places throughout the West in both national parks and national forests where deer and elk have been protected and their enemies destroyed."[49] This became the common wisdom.

Game managers also had to work for unity among wildlife advocates. Though they emphasized "a shootable surplus," managers needed the political support of the larger public. When a critic attacked his American Game Plan of 1930 as something for the man who just wanted to kill more ducks, Leopold pointed to the program's wider goals. The American Game Plan was designed for the country as it was, interested in hunting and intent on making things "pay." But, he promised, it would also make wildlife "and the enjoyment of it a part of the normal environment of every boy, whether he live next door to a public sanctuary or not." He pleaded for sportsmen and nature lovers to hang together to save wildlife. "At present," he warned, "we are getting good and ready to hang separately."[50]

The development of game management had important effects on the Biological Survey. Aside from the PARC program the agency had few continuing responsibilities that would gain it public support, and some Survey workers saw in game management a mission for the bureau and a place for themselves. W. L. McAtee was the most prominent. He had started with the Survey in 1904, analyzing the stomach contents of birds, and had worked his way up to head the Division of Food Habits Research. He had been unhappy for some years about what he saw as the Survey's gradual loss of scientific respectability and its growing commitment to predator control. When he lost his position in a reorganization in 1934 (for which he blamed Clarence Cottam), he made a new place for himself as a source for information on game research. In 1935 McAtee convinced his superiors to support a regular bibliographic review of research on game.[51] He helped form the Wildlife Society in 1936, was one of its two trustees, and became the first editor of the *Journal of Wildlife Management* two years later.[52] Survey chief Ira Gabrielson sent a letter to all employees urging them to join. Many did. Survey workers constituted at least 10 percent of the members, and the agency had representatives on virtually every committee.[53]

The Survey and game management had a symbiotic relationship in other ways as well. The agency had been responsible for the administration of federal wildlife programs, and as the New Deal funded research it gave the Survey the task of managing it. In 1935 the agency began overseeing the work of a new set of cooperative wildlife research stations. They were tied to land-grant colleges and to state conservation agencies (the Wisconsin Department of Conservation refused to cooperate, a disappointment for Leopold), and constituted a new avenue of influence for federal workers. Three years later, when Congress provided federal funding of state game-management research under the Pittman-Robertson Act, the Survey was to set and enforce standards. Federal reorganization plans amalgamated it with the Bureau of Fisheries, changed its name to the United States Fish and Wildlife Service, and put it in the Department of the Interior in 1939–1940, but by then the change was complete. The agency was a national wildlife research organization, and it could call on an established discipline of game management to support it.[54]

New Policy in Old Bureaus

Game managers were not the only group catering to public interest in wildlife, nor was the Survey the only agency helping to spread new ideas about wildlife. The rising public interest in wildlife and the New Deal's interest in wildlife conservation led to the diffusion of wildlife work into a number of agencies. The National Park Service was most affected, and it, in turn, had the greatest impact on the public. Wildlife protection had been one of its legal responsibilities from the time the first parks were set up, but not all animals were protected. Park managers, and the Park Service after it was formed in 1916, had encouraged the "beautiful" species that the tourists wanted to see, such as elk and deer. They neglected others—except in the case of predators, which they often killed to "save" the "innocent" creatures.[55] In 1926, in response to the mammalogists' campaign against the Survey's poisoning policies, Park Service director Stephen Mather declared that it was not park policy to exterminate any native animal (though he did endorse predator control to protect the "weaker" species). Two years later the conference of park superintendents formally condemned predator control in the parks. In 1931, Mather's successor, Horace Albright, banned poison in the parks, partly in response to the Howell petition. The Park Service, he declared, would "give total protection to all animal life."[56] Control, even killing, of individual animals that threatened human life or property continued, as it does today, but predator control as a way of increasing the numbers of "nice" an-

79

imals was on the way out. In 1935 the Park Service ended the last predator control operation outside Alaska—coyote trapping in Yellowstone—although it is noteworthy that some fifteen wolves were killed in a control operation on a sheep range in Mt. McKinley National Park between 1945 and 1951.[57]

The Park Service also began the scientific study of wildlife in the parks. In 1928 George Wright, a part-time ranger, convinced the Service to begin a two-year survey of wildlife on its land (his offer to pay for the project may have affected the decision). The results were so impressive that the agency appointed him a field naturalist in 1931 and made him chief of the new Wildlife Division two years later. The Division promptly published the original faunal survey as the first of a new set of monographs, Fauna of the National Parks.[58] It was more than a report; it called for a new park policy. All species should be saved. Each "is the embodied story of natural forces which have been operative for millions of years and is therefore a priceless creation, a living embodiment of the past."[59] Protection was particularly important for the larger predators. In the parks they "find their only sure haven . . . [and] are given opportunity to forget that man is the implacable enemy of their kind, so that they lose their fear and submit to close scrutiny."[60] Park management should move from an "urban to a wilderness concept of the preservation of wildlife" and educate visitors to seek out bears in their natural surroundings, not at garbage dumps.[61]

The Wildlife Division set in motion a policy and a research program that would have important consequences for our view of nature. It funded and then published Adolph Murie's work on coyotes and wolves. It supported the move to make Isle Royale a national park, preserved as a wilderness area. When wolves colonized the island in the early 1950s the Park Service welcomed and studied them. In 1957 it contracted with Durward Allen to study the wolves and moose. This project was the training ground for the current chairman of the Wolf Specialist Group of the Species Survival Commission of the International Union for the Conservation of Nature, L. David Mech. The wolves of Isle Royale, the only ones in the United States outside Alaska to live under natural conditions, are still the subject of continuing scientific research.[62]

The Park Service was uniquely placed to carry out a policy of predator preservation. Game managers might preach an end to indiscriminate killing, but their professional aim was a larger game crop. Where predators interfered they had to go. The Park Service had the mission of protecting and preserving all wildlife. It was charged with keeping wild nature wild. There were, in the 1930s, no wolves in the parks, but the Service pressed ahead with protection of other wild animals. With the cooperation of the Biological Survey and Congress, which set aside a refuge (Red Lakes Migratory Waterfowl Refuge), the Division began work to save the trumpeter swan. The Park Service also began to protect the grizzlies in Yellowstone; it worked with the states to preserve the bighorn; and it shot out the burros in the Grand Canyon to provide more food for the native animals.[63]

The Park Service could also influence and educate the public in ways other agencies could not. Naturalist talks attracted a large and interested audience that already knew something about nature and would be receptive to new ideas. Like wildlife research, nature education blossomed in the 1930s. It had started in 1918, when Harold Bryant, one of Grinnell's students, began giving nature lectures in Yosemite National Park. Bryant and others expanded the program during the 1920s, largely with volunteer labor, and in 1930 the Service brought him to Washington to organize a program for all the parks. He had been on the advisory committee that had recommended creating the Wildlife Division, and he was committed to the goals Wright had recommended in the original faunal survey. Park education thus meshed with scientific research.[64]

Change came slowly, for many resisted the idea of natural controls on park animal populations. Predators, they thought, would eat up the animals the tourists wanted to see. Retired Park Service director Horace Albright wrote angry letters in 1937 and 1938, condemning the continuing study of coyotes. The animals, he thought, should be reduced or eliminated. Visitors never saw them, and they were not needed. Stockowners around the parks were equally angry, particularly those that might be affected by a larger grizzly population in Yellowstone.[65] The public also wanted "outdoor zoos." The result was a mixed bag. Reporting in 1940 to the Senate's special committee on wildlife, the Service called for a new appreciation of wildlife in the context of the parks, for basic ecological research on predators and prey, and for the preservation of all species in a natural balance. At the same time it said that in 1939 "the daily 'grizzly's banquet' at Canyon was attended by 106,615 persons, who benefited by a close yet safe acquaintance with the animals and by hearing of conservation methods and ideals from the attendant ranger-naturalists."[66]

Wildlife policy found a place in other agencies as well. The Roosevelt administration rationalized existing programs and agencies, expanded the scope of their operations, and made wildlife preservation an integral part of conservation programs. Harold Ickes, Roosevelt's conservation-minded Secretary of the Interior, remade that agency, turning it from a catch-all group with a reputation for inefficiency and corruption into something resembling a department of conservation and nature protection. He failed in his grand design to put all conservation and natural-resources work in his agency (the Forest Service and its allies fought him off), but he did get wildlife.

Wildlife preservation was made part of the mission of the Civilian Conservation Corps and the Soil Conservation Service. The CCC and SCS programs changed the conditions on millions of acres, gave farmers and wildlife managers experience in game production, and helped make wildlife work a normal part of conservation. The agencies also provided an important source of jobs for the new professional wildlife scientists. The addition of wildlife research stations to agricultural experiment stations brought together academic research, state conservation departments, and federal workers. Wildlife bulletins and wildlife talks on extension-service radio programs reached millions of Americans.

Legislators made other changes. The Wildlife Coordination Act of 1934 required federal agencies to consider the impact of their actions on wildlife—though it did make that subordinate to their primary missions. The Forest Wildlife Refuge Act gave the Forest Service statutory authority and a mission to protect game on its lands. The Duck Stamp Act of 1934 taxed duck hunters to create a fund for waterfowl refuges, and the Pittman-Robertson Act of 1938 did much the same for wildlife research. A tax on firearms and ammunition went toward state wildlife research work.[67]

World War II ended reform efforts and drastically reduced many domestic programs, but by then the scientists, administrators, and legislators had laid the foundations for an active wildlife preservation policy. Goals had been set. Agencies, research programs, and research stations had been established. A professional corps of workers had formed a self-conscious scientific community, and a new science justified protection of all species and saw a place for each one. Olaus Murie had suggested that mammalogists had not thought through their commitment to the coyote. Now they could, using the intellectual concepts of animal ecology. They would, and they would bring the public along with them. Even as federal work went on, the ideas of animal ecology had begun to diffuse outward from the scientists to the public, where they would reshape nature literature and, eventually, public perceptions of wildlife and nature.

CHAPTER SIX

From Knowing to Feeling
Changing Ideas about Nature,
1925-1950

On those rare occasions when I think back to the climate of opinion that surrounded zoologists in the 1930s I realize that few of us foresaw the emergence of two public concerns which now loom in our vision and which govern our actions. These are respect for nature and naturalness (the environmental-ecological idea) and respect for life itself (the anti-killing idea). In the 1930s we supposed that we could "manage" nature and "improve" or "reclaim" the earth's ancient, time-tested organic systems. Now we are less sure.—Victor Sheffer, *Adventures of a Zoologist* (1980)

Though ecological and humane ideas both "emerged" in the 1930s and went on to "loom in our vision and . . . govern our actions" three decades later, their progress was quite different. Revulsion at needless suffering, including the suffering of animals, and the belief that all creatures had some mental life had been growing in Western societies since the seventeenth century. In England and America, where these ideas were strongest, it was a natural development to extend them to wild animals, which Americans did as they became more interested in nature and as the numbers of draft animals declined.[1] Ecology was a different matter. It required a fundamental change in the way people looked at nature. It is not a mistake, though, to link humane concerns and ecology in the interwar period. Both owed something to the growing public interest in wildlife, both began to affect public sentiment about wildlife during these years, and both laid the groundwork for their participation in the wildlife politics during the "environmental era."

From Appreciation to Ethics

Humane and ecological concerns developed in the context of a growing public interest in wildlife. Despite the Depression, people continued to visit the national parks and camp in the national forests. They also bought nature books: Roger Tory Peterson's *Field Guide to the Birds* was a phenomenal success in 1934. They wanted wildlife as an attraction of the national parks. Everglades, authorized in 1934, lacked the scenic grandeur associated with the parks—and some conservatives opposed its inclusion in the system on that ground—but it was a popular addition to the system.[2] That most sensitive of indicators, elected officials, had begun to respond even before Franklin Roosevelt made conservation and nature a part of New Deal programs. The Senate formed a Special Committee on the Conservation of Wildlife Resources in 1930; the House followed suit in 1934.

The first people to change in response to the new perspective coming from ecology were those already intensely interested in wildlife. The Audubon Society began to shift its policies and educate its members in the 1930s. Earlier it had defended birds by arguing that they were useful to man. Until beauty will "be a sufficient reason" to protect birds, a 1900 *Bird-Lore* editorial declared, "we must base our appeals . . . on more material grounds."[3] Science should "hold the scales" (by "science" it meant the Bureau of Biological Survey's work on food habits). It did not shrink from the necessary conclusion. A "verdict in the birds' favor cannot always be expected"; some will "fall under the ban."[4] The Society's only argument against a general policy of shooting "bad" birds—mainly the bird-eating hawks—was that most people could not tell them apart and would shoot all raptors. On refuges, where shooting would be done by experts, it did argue that hawks and owls should be killed.[5]

Later the Society suggested that even the bird-killing hawks be preserved. They were a heritage of the past, part of the record of evolution.[6] In 1931 *Bird-Lore* published W. L. McAtee's "A Little Essay on 'Vermin.' " All species, the Survey's expert argued, were admirable, and all should be saved. In 1935 the Society really changed course. An article on "Feathered and Human Predators" gave prominent place to Errington's work on the bobwhite. Another reported Leopold's remarks to the American Game Conference on the beauty of birds and the futility of most predator control work (Leopold became a director of the Audubon Society in 1935).[7] Audubon backed away from "vermin" control on bird preserves. Shooting or trapping, it said, should be done selectively—and only when scientific study showed it was needed.[8]

More was at stake than the reputation of hawks or the definition of a bird sanctuary. Ecology suggested that it would be impossible to "save" a single species. Each depended on, each lived in, a complex world. Only by saving that world in all its complexity could nature be preserved. The Audubon Society would not follow that logic to its conclusion until the 1960s, but by the late 1930s it had begun to travel that road. It became less concerned about hunting seasons, more interested in habitat preservation. It paid less attention to urban bird sanctuaries and breeding grounds, more to areas that provided for all of a species' requirements. Spectacular birds were no longer the center of attention; the Society wanted to save *all* kinds. It also expanded its efforts to take in mammals and other forms of wildlife.

While the Audubon Society was teaching its members to appreciate all forms of life as part of an interdependent world, nature writers were reaching a larger public. Forty years earlier Seton and Roberts had helped people find sublimity in the world of Darwinian nature. Now a new generation showed the beauty of the "web of life" that connected all of nature, making a harmonious whole. They gave scientific warrant to the Romantic conception of a world in which each part had a purpose and a place. They fostered an emotional identification with nature through a vision of order based on the new science.

One of the first writers to incorporate this perspective was Donald Culross Peattie, a botanist who worked for the Department of Agriculture. He was already a recognized nature writer in 1938 when he published *A Prairie Grove*, which Aldo Leopold hailed as the "first ecological novel."[9] It appears to be part of the literature of pioneer nostalgia, the old tale of man settling and conquering the new land. It is not. Peattie was not concerned with the deerslayer or the conflict of natural man with civilization. His story was "the island grove, the trees and the great grass, wildfowl and the furred and antlered beasts, and tall men, very small, moving about in their roles beneath lofty boughs and across wide spaces."[10] The book was not a "novel, not a historical romance, not a popularization of history. I say that I am remembering, remembering for the trees and the great grass province and the passenger pigeons and the wild swans."[11] It was an extended reflection, using the techniques of the novel and the nature essay, on the changes white men made on the land that is now Illinois.

It is not the frontier and its virtues that are lost, but the world of nature, and it is gone forever. "Do you want back Eden?" he asks. You cannot have it. "You cannot have it both ways. You cannot call yourself a man and go back to the wolf's way. Our way has gone forth everywhere."[12] He ends with the migration of the last passenger pigeons and the turn of the year. "The seed is in, fateful and indomitable; we have populated where we have slain. Still sometimes when in fall or spring the wind turns, coming from a fresh place, we smell wilderness on it, and this is heartbreak and delight."[13] This is also Romanticism— but it is Romanticism with a difference. Here there is no sublime scenery, no noble savage, no natural man. What inspires awe is the complex world of nature that settlement destroyed, and that world is seen by science. Peattie, in fact, appended a bibliography of scientific studies he had used to re-create the prairie province. It included the classic text on prairie ecology, Frederick Clements's *Plant Succession.*

Another writer who used the new science was marine biologist Rachel Carson. Her first book, *Under the Sea Wind*, appeared in 1941.[14] Place it beside a work published only thirteen years earlier, one Carson herself admired—Henry Beston's *The Outermost House.*[15] Both employ a classic form, the chronicle of a year. Both have as subject the life of the Atlantic coast. But there the resemblance ends. Beston's is a personal narrative of a year he spent in a cottage far out on Cape Cod. We see the shore, the wildlife, the Coast Guardsmen who patrol the beach, the changing seasons, and the cycle of nature, but we see them through his eyes and as part of his experience.

Carson dispenses with time, place, and human lives. There is no named observer, no human point of view, no lessons of life, no person living in nature. There is only the Atlantic coast, the seasons, and the life of the sea. Sections of the book describe the Carolina coast in the spring and summer (with a digression to follow a sanderling to the Arctic for the nesting season), the growth and life of a mackerel, and the life cycle of the eel. She emphasizes the processes of nature and the food chains that bind the creatures together as part of a system. People enter only casually and accidentally. They are part of or disturbers of the life of the sea. This is not nature as Carson has seen it but as she has constructed it from scientific study.

Her purpose, she said, was to make the sea and its creatures "as vivid a reality for those who may read the book as it has become for me during the past decade . . . out of the deep conviction that the life of the sea is worth knowing." To appreciate its life and movement "is to have knowledge of things that are as nearly eternal as any earthly life can be."[16] Like *The Outermost House, Under the Sea Wind* is a Romantic statement that tries to induce in the reader a feeling of awe in the face of nature. But it uses ecology, not personal observation and reactions.

Even writers without formal scientific training took up the new ideas. Sally Carrighar's popular *One Day at Beetle Rock* (1944) is an example. The form was one familiar in nature writing— the intense study of a small area—but the perspective was new.[17] Carrighar presented the life of the area around Beetle Rock on a single day through a series of linked chapters. Each dealt with a different animal, and incidents recurred as the viewpoint changed. We experience a chase from the point of view of hunter and hunted, and a storm as several animals see it. The center of the book, the common thread, is the web of relationships that make up the life at Beetle Rock.

Seeing nature as a web of interdependent organisms linked by food chains connected in trophic levels made predation a neutral phenomenon—one of the mechanisms that kept natural populations in balance. Predators were now not evil, even a necessary evil; they were part of the "web of life." Their presence was not disturbing but evidence of the completeness of that part of the natural world. This would help change the popular image of the wolf, or provide justification for a change that was already in progress. As wilderness became scarce and precious, as it lost its aura of danger and became a retreat from civilization, the prime wilderness predator, the wolf, acquired a new image. In the early 1940s John Stanwell-Fletcher, a Canadian naturalist, could call the time he and his wife spent in the back country of British Columbia "Three Years in the Wolves' Wilderness." They, "perhaps more than any other creatures, seemed to us to be the spirit of the wilderness itself." It was a spirit and a presence that should be saved. "If the wolf is exterminated," he said, "we shall have lost one of the most virile, wise, and beautiful of all wild creatures."[18] This is much like Seton's attitude, but for the Stanwell-Fletchers the wilderness is not something that will be developed; it is something that ought to be preserved. The wolf is not to be destroyed; it can and should be saved.[19] Others, at least a few, had reached the same conclusion. In 1946 Victor Cahalane, biologist for the National Park Service, asked the Wilderness Society, "Shall We Save the Larger Carnivores?"[20]

One person who was even then coming to the conclusion that we indeed should save the larger carnivores was Aldo Leopold. In his case, though, the approach was not through study of wolves, nor even by way of wilderness. Though he was deeply interested in wilderness and its place in American culture, Leopold's field experience and research—which powerfully shaped his ideas—were with farm and forest game, animals that lived comfortably near humans. For that reason the path he traced from killing wolves to calling for their protection is all the more remarkable.[21]

Leopold grew up with conventional ideas about animals and nature. Game species were the only important ones, and they were to be produced like any other crop. Predators were like weeds or noxious insects. They should be eliminated. His training in forestry (he graduated from Yale Forestry School in 1909) reinforced this perspective and in his first years on the job he acted by these convictions. As a Forest Service ranger in New Mexico (1910–1918) he cooperated with the Biological Survey's predator control efforts. In 1919, editing *The Pine Cone* for the New Mexico Game Protective Association, he urged "going out after the last lion scalp, and getting it." He shot wolves whenever and wherever he could find them.[22]

"I was young then," he later wrote, "and full of trigger-itch; I thought that because fewer wolves meant more deer, that no wolves would mean hunters' paradise."[23] Speaking of predator control he wrote: "At the time I sensed only a vague uneasiness about the ethics of this action. It required the unfolding of official 'predator control' through two decades finally to convince me that I had helped to extirpate the grizzly from the Southwest, and thus played the role of accessory in an ecological murder. . . . As a boy, I had read, with intense sympathy, Seton's masterly biography of a lobo wolf, but I nevertheless was able to rationalize the extermination of the wolf by calling it deer management. I had to learn the hard way that excessive multiplication is a far deadlier enemy to deer than any wolf."[24]

It was more than the "unfolding of official predator control" or the "excessive multiplication" of deer that changed Leopold's mind. He came to see these phenomena, in fact, within a perspective shaped by his own experiences and by contact with people with other ideas and other ways of looking at nature. The first group, perhaps, was the professional mammalogists and ornithologists he began to communicate with around 1917, when he began writing up field observations from his hunting expeditions as notes for scientific journals. One of them was *Condor*, and this brought him into contact with chief editor Joseph Grinnell and others on the staff of the Museum of Vertebrate

Zoology in Berkeley. They had a different interest in wildlife and helped change the forester's ideas. In 1922, for instance, Leopold shot a roadrunner to see what it had in its mouth. His son Starker recalled that his father became quite excited when he found that it was a quail chick. He must have been, for the note he fired off to *Condor* describing the incident ended with the observation that he had never before killed a roadrunner, but they are "now on my 'blacklist' and will stay there until somebody proves that this was an exception to their usual habits." Associate editor Harry Swarth suggested that this was extreme punishment for the "occasional incident, for it seems to me that the roadrunner is just as desirable a bird to have around as the quail!" "On thinking the matter over," Leopold replied, "I believe I was a bit hasty in inserting the last paragraph of my article. As a usual thing, I am heartily in sympathy with your viewpoint." He asked Swarth to delete the reference to blacklisting.[25]

In 1924 Leopold left the Southwest to take up a new position as assistant head of the Forest Service's Forest Products Laboratory in Madison, Wisconsin. Leopold later claimed that the "industrial motif of this otherwise admirable institution was so little to my liking that I was moved to set down my naturalistic philosophy" in essays.[26] This is stretching it; he had already begun to react to what he called "boosterism" while still in the Southwest.[27] The transfer did remove him from the line just as the Kaibab disaster was reaching its peak. It gave him the luxury of facing the problem and examining it without having to justify and defend policy. He was drifting away from the Forest Service and its goals, and in 1928 resigned to make his way as an independent game consultant. An intensive self-education in game and the conditions necessary to produce it followed. He did a survey of the Midwest for the Sporting Arms and Ammunition Manufacturers' Institute. He began a campaign to make game management an applied science.[28] He got a temporary position teaching game management (the first one in the country) in the agricultural economics department at the University of Wisconsin, and began putting students to work on game-management projects. In 1931 he met the English ecologist Charles Elton and became an enthusiastic advocate of his ideas and of his book, *Animal Ecology*.[29]

By the early 1930s Leopold was moving beyond the idea of game as a crop. Even as he wrote *Game Management* he was discarding its utilitarian approach and coming to think of all wildlife as part of a natural system. Management came to seem less a matter of producing a crop than of preserving the system on the land.[30] His students' research and Elton's ideas now came to bear directly on the question of deer. The Kaibab and other irruptions forced him to rethink his position. It was, apparently, not enough to remove the checks that held down deer reproduction. The species was not self-regulating. Nor could managers rely on human hunters to trim the herds. In the mid-1930s Leopold made two trips that helped crystallize his thinking. In Germany he saw a completely artificial system. The Germans were tree farming and deer farming. The forests consisted of large stands of the same species, all the same age. The herds were maintained in this sterile environment by feeding. In Mexico, on the Rio Gavilan, he found the polar opposite, a deer herd living under natural conditions, controlled by natural forces. The contrast confirmed what was becoming his credo: the best management was that which restored or preserved natural mechanisms of control.[31]

His experiences as a Wisconsin game commissioner in the early 1940s added a personal dimension. The deer herd in the northern part of the state had grown rapidly in the 1930s, fattening on new growth on logged-off tracts and abandoned farm land. As the brush and low trees gave way to forest, the food supply went down and the pattern of the Kaibab recurred. The deer herd crashed; the range was damaged; and the weakened survivors fell prey to disease. Leopold argued that the herd and the range could be saved only by killing more deer. As in Arizona, hunters opposed this idea. So did resort owners, fearful of losing the attraction of large numbers of deer. "Bambi" had just reached the movie screen (1942), and many people protested against the "slaughter" of the "beautiful" deer. Leopold wrote, spoke, and organized tours of the "yards" where newspapermen and humane advocates could see the starving deer for themselves, but it was an uphill battle. Led by hunters who sneered at the "book learning" of "professor Leopold," opponents of an increased kill pressed the legislature and the governor to "save our deer."[32]

This may have reinforced Leopold's conviction that natural controls were best. Wolves, unlike hunters, did not write letters to the editor attacking game management practices. It was at most reinforcement. Leopold was sure that human management was not as good as a functioning natural system, and he was already making his ecological insights into moral principle. "A thing is right," he said in his essay, "The

Land Ethic," "when it tends to preserve the integrity, the stability, and beauty of the biotic community. It is wrong when it tends otherwise."[33] This was, he recognized, an act of faith. Ecology could suggest it but not prove it. In 1939, commenting on the relationship of coyotes to rodent populations (that topic again!), he said quite frankly that no one really knew what was going on. What was needed was a synthesis of recent research. He doubted his "own ability to do it, both in point of ability and in point of time," and hoped that Errington would do the job.[34] Murie's study of the wolves and sheep of Mt. McKinley bolstered that faith. In a long memo Leopold pointed out that Murie had found that wolves and sheep formed a self-regulating system. A healthy predator population prevented irruptions. When sheep populations were low the animals could all live in the rough areas of the range, secure from wolf predation. When the population was high, some had to live in less favorable areas and the wolves got more of them. *The Wolves of Mt. McKinley*, he concluded, was an important contribution to a sound conservation policy.[35]

At the end of Leopold's intellectual journey was the wolf, which had now became a symbol of a wilderness that should be preserved and which he now saw as an essential part of the system of nature. In "Thinking Like a Mountain" (1944) he presented his mature thoughts on what he saw as a linked set—deer, wolves, and forests. "Only the mountain has lived long enough to listen objectively to the howl of the wolf." Only the mountain knows the "deeper meaning." We all know the meaning is there, "for it is felt in all wolf country, and distinguishes that country from all other land." His own conviction came from the day he and his friends killed a wolf on a hunting trip. Watching the "fierce green fire dying in [the wolf's] eyes," he sensed that his own simple ideas were wrong. Since then "I have lived to see state after state extirpate its wolves," and watched the deer multiply and destroy the range. "I now suspect that just as a deer herd lives in mortal fear of its wolves, so does a mountain live in mortal fear of its deer."[36]

Cruelty and Wildlife, 1920-1938

Ecological ideas were not the common currency of nature lovers during the Depression years, though. Leopold was well in advance of his contemporaries. So too, it now seems, were the lonely pioneers seeking to extend the ideals of humane treatment to wildlife. Today humane advocates mount well-financed campaigns against trapping and hunting, and they enjoy considerable popular support. In the interwar period the movement was much smaller, poorer, and less popular. It was only a shadow of what was to come.

In the 1920s wildlife became a significant part of the humane movement, with the galvanizing issue being the common steel leg-hold trap.[37] The Society for the Prevention of Cruelty to Animals and the American Humane Association offered rewards for humane traps, the AHA offered bounties for box traps, painless leg-hold traps, and instant-kill traps.[38] The Anti-Steel Trap League did more. Its name was its program, and between 1925, when it was formed, and 1942, when World War II ended its efforts, it sponsored or wrote at least ninety-nine anti-trapping measures introduced in eighteen states. Five states and twenty municipalities either banned trapping or the leg-hold trap. Where it could not get a ban it sought other, lesser measures, with considerable success. Thirty-seven states passed new trapping laws that ranged from new regulations on seasons to requirements that trappers check their lines at regular intervals.[39]

Like hunting and nature appreciation, game management and the anti-trap movement drew gender lines. Men did the managing, women the protesting. Women, said Edward Breck, president of the Anti-Steel Trap League, were "the moral backbone of the land, and it is primarily their chief duty to right this great wrong, just as they have been the chief offenders in condoning it, though innocent, for the most part, through ignorance."[40] The League sought to dispel that ignorance. Each fur coat, it proclaimed, had been purchased at the cost of the suffering and death of many innocent animals (they were, to be sure, predatory animals, but the humane movement now believed even they should not suffer). Anti-trap advocates appealed not only to sympathy but to religion, developing the arguments the Reverend William Long had offered in "Animal Immortality" into what one historian called "ecumenical visions of Christ's mercy and Nature's virtue."

> Then through the gloom that night came ONE
> Who set the timid spirit free:
> "I know thine anguish, little son—
> So once men trapped and tortured ME."

Thus ended "To A Fur Scarf," a poem the League used in its pamphlets.[41]

The League did not stop with education, persuasion, or poetry. It wanted laws, and to get them it formed coalitions and lobbied legislatures. It relied on elite and middle-class women, mobilized by women's clubs. There was no particular class appeal, but poor women did not buy fur coats. It also took help wherever it could find it, from groups as diverse as the Massachusetts Rod and Gun Club and the Ohio Academy of Sciences.

Kentucky's ban on leg-hold traps is a good example of the organization, zeal, and practical politics that marked the League's work. The state leader was Miss Lucy Furman, a retired schoolteacher and novelist, who had written an eloquent article on the cruelty of trapping in 1928. It had attracted the League's attention, and it made her a vice president. She set out to convert Kentucky to the cause. Her major obstacle was the trappers' claims that there was no alternative to the usual trap. To pass the League's bill would abolish trapping altogether, and legislators were unwilling to do that. In 1933 Furman got help from an unexpected quarter: Vernon Bailey patented a humane animal trap. Instead of rigid steel jaws his had a chain circle. When the trap was sprung, the chain contracted to make a tight but flexible loop around the animal's foreleg.[42] Furman returned to Kentucky from her retirement home in Florida, triumphantly bearing aloft (metaphorically speaking) the "Verbail" trap. She made it the centerpiece of her campaign. In 1936 she enlisted the Kentucky Federation of Women's Clubs. Two years later she convinced the country's largest trap manufacturer, the Animal Trap Company of America, to manufacture Bailey's device. The Kentucky legislature passed her bill in 1940.[43]

Compromise marked all the League's efforts. It did not stand against trapping or animal killing. It wanted it done with a minimum of pain. Like the Audubon Society's defense of birds on economic grounds, this was a strategic judgment, not a position of principle. The public would not accept a ban on trapping, and half a loaf was better than none. There were, inevitably, those who held the opposite view. One was Rosalie (Mrs. C. N.) Edge, whose Emergency Conservation Committee was for a decade a strong, but divisive, force for animal welfare and protection. She was the founder and sole propri-

etor from the committee's beginning in 1929 to its demise in the late
1930s. Her activity was the most visible manifestation of the radical
wing of the humane movement, but the issues she raised continued to
be problems for wildlife's defenders.[44]

Her main target was the leadership of the Audubon Society. It had,
she charged, sold out. T. Gilbert Pearson, secretary since 1910 and
president since 1924, had taken money from the arms companies and
profited from the sale of Audubon memberships. He had dragged his
feet on creating more wildlife refuges, had allowed trapping on Audubon sanctuaries, and supported high bag limits on ducks. The directors, a self-perpetuating clique, also had to be replaced. Edge did
not pursue her quest for justice by the gentlemanly methods favored
by Audubon's directors. The Emergency Conservation Committee
published a pamphlet about trapping on the Audubon sanctuaries,
"The Audubon Steel-Trapping Sanctuary." The subtitle was: "The Directors Condone Cruelty for Profit." The text was no milder and her
other activities just as aggressive. She asked the Audubon Society for
its mailing lists so she could make her case to all the members. When
it refused she sued. When her candidates were not elected to the
board she went back to court, claiming irregularities in the voting.[45]

Edge also attacked the Biological Survey's predator poisoning campaign. The Committee's pamphlet on this subject was called "The
United States Bureau of Destruction and Extermination: The Misnamed and Perverted 'Biological Survey.'" It charged that predator
control work was destroying native wildlife. Probably relying on C. C.
Adams and possibly on W. L. McAtee, it said that the Survey had
abandoned scientific research in favor of abject subservience to the
wishes of the western livestock industry. It reprinted, with all the signatures, Howell's petition of 1931.[46]

These stands alienated many people, but Edge had strong, if limited, support. The Anti-Steel Trap League published its own pamphlet about trapping on Audubon's sanctuaries, "Blood Money for the
Audubon Association."[47] McAtee privately confessed to a strong temp-
tation to make a contribution. Others, less tied by their official positions, did. C. C. Adams was on her board of consultants. William Hornaday, one of the most aggressive defenders of wildlife (and a
reformed trophy hunter) applauded her work.[48]

The problems Edge raised were not new. In 1911 the Audubon Society had split over the propriety of taking money from arms and ammunition manufacturers. Its campaign against collecting eggs and shooting songbirds had roused fears in the scientific community that public pressure would hinder scientific activities.[49] There had always been tension between people primarily interested in appreciating nature and those who wanted to do something else. It was heightened by the ties among the groups. Hunters were the main political and financial support of wildlife preservation programs. Nature enthusiasts and humane crusaders for wildlife were relying on people with whom they disagreed. Even the scientists occasionally found themselves in trouble, for they collected songbirds for museums. Edge's militance was a threat to an alliance that all but the most extreme accepted as necessary.

So long as Audubon did not take up Edge's crusade, the Survey had no reason to listen to her. It did not, dismissing her as a "sentimentalist" meddling in "practical" affairs she did not understand. The Society had more difficulty. As a member Edge had a right to be heard. The directors tried a combination of compromise, counterargument, and strategic retreat. They ended trapping on the refuges but fought off efforts to unseat Pearson and the directors. By 1938, when her lawyer withdrew from the case (with evident relief), claiming that the major disputes had been settled, Edge seemed almost resigned. The directors now treated her, she told the attorney, with respect and kid gloves. They were not, though, going to change. They were too afraid of losing their tax exemption.[50]

EDGE'S CAMPAIGN failed, and the Anti-Steel Trap League faded in the early years of the war. The issues did not. Americans continued to be interested in nature and more and more were committed to the "environmental-ecological idea and the anti-killing idea." These two ideas, however, were at odds. One valued the stability of the system. It thought in terms of species and populations, not individuals, and accepted animal death and hunting. The other was opposed, at least in principle, to all killing. It placed individuals and their suffering first. There were as yet few who would push the logic of humane treatment to its extreme—the moral obligation to avoid inflicting any pain on any creature—but that would come. It would come quietly. The gospel of ecology was more obvious and influential. Nature literature had begun to spread that gospel. After World War II public experience with environmental problems and education, often by way of movies and television, would lead the public to wonder about ecosystems, the connections within them, and man's effects on the "web of life." By the 1960s many things would appear in a new light.

CHAPTER SEVEN

The Public and Ecology

1945-1968

In 1948 Alan Devoe wrote:

[I]n a much more than poetic sense, we are brothers of the whooping cranes.
. . . The fact of our brotherhood is zoological fact, ecological fact, everlasting
earth-fact; and we are to forget it only at our grave peril. [When the] web of
nature's being is anywhere broken or seriously disturbed, the effect travels
throughout the whole of things, man not excepted.[1]

At that time, his was a minority view, even among wildlife lovers.
Twenty years later his sentiments were common. People had looked at
nature as an unlimited source of unconnected resources or objects and
believed that human actions were, by and large, unimportant. They
saw animals as part of the landscape, objects for appreciation or pursuit. By the late 1960s many people viewed nature as a complex, frag-
ile web, of which animals, all animals, were integral parts. They believed humans could easily destroy nature by uncontrolled industrial
development. They demonstrated their commitment in votes, rallies,
and petitions. They demanded government action to protect the environment and preserve ecosystems.

These ideas became popular because the public learned new things
about nature, including a new way of looking at it, and because their
experiences reinforced that vision. The booming postwar economy
gave middle-class Americans more money and more leisure time. It
drew them from the country and small towns into the cities. They had
less opportunity to see nature in their daily lives but more chances to
take vacations in the wild. Their culture valued "contact with nature,"
and this, combined with their new opportunities, brought people to
national parks and forests and the woods, fields, and streams of America in unprecedented numbers. There they found that economic development and population pressure were causing more polluted
streams and dirty air and destroying forests and wilderness at a high
and rising pace. "Nature" was vanishing, and people were responsible.

They learned about less obvious effects from books, magazines, movies, and television. Even those with no serious interest in the outdoors found the connections in nature and the problems of human contamination of the environment of more than theoretical interest. Fallout from atomic bombs put radioactive isotopes in the milk children drank for breakfast. New pesticides left residues that might cause cancer on fruits and vegetables and in meat. That all things were connected in a vast whole came to seem less a Romantic vision than an alarming reality.

The rapid pace of development revived Malthusian fears of shortages and eventual starvation. Books like William Vogt's *Road to Survival* and Fairfield Osborn's *Our Plundered Planet*, both published in 1948, said that soil erosion, the cutting down of vast tracts of forest lands, human population growth, and the rapid depletion of natural resources were threatening our future.[2] Our current policies, Vogt warned, particularly our neglect of the soil, were destroying the base on which our civilization rested. Osborn agreed. He had written *Our Plundered Planet*, he said, to "show that if we continue to disregard nature and its principles the days of our civilization are numbered."[3]

Attacks on waste and improvidence were not new. These, however, had a new perspective. Ecology now supported conservation and preservation. Men, Osborn said, had always "been intrigued by the idea of the harmony of nature and the vision of man working in harmony with that vast symphony." We now know "the concept of poets and philosophers in earlier times is a reality. . . . Each part is dependent upon another, all are related to the movement of the whole."[4] Vogt argued for "ecological health" as a goal of management. He discussed natural resource problems in terms of "carrying capacity," "biotic potential," and "environmental resistance"—new terms for a new situation. After reading the manuscript of *Road to Survival*, Leopold told Vogt that "[t]he only thing you have left out is whether the philosophy of industrial culture is not, in its ultimate development, irreconcilable with ecological conservation. I think it is."[5]

There were less theoretical responses. Wildlife scientists had been concerned about African and Asian game herds since the early 1900s, and had tried in the 1930s to organize to save the animals. In 1948 they did. A group formed the International Union for the Conservation of Nature and Natural Resources (IUCN). Two things distinguished it from earlier movements. One was its extensive international connections. It drew on scientists from several countries and was organized under the auspices of the United Nations Educational, Scientific and Cultural Organization (UNESCO). The second change was the new group's concern with extinction. It concentrated as no one had done before on accumulating information on species that were in danger of vanishing, and it became the standard source on endangered species.[6]

Wildlife and nature organizations reflected these new concerns. There was a strong and growing interest in the preservation of wilderness and wild areas as wildlife habitat. The Wilderness Society had been an anomaly when it had been formed in 1935. It soon blended in. Both the Conservation Foundation (founded 1948) and the Nature Conservancy (1950) were set up to save areas in their natural state. The Sierra Club, just becoming a national organization, began sponsoring annual conferences on wilderness in 1949, and in the early 1950s it dropped from its bylaws the phrase about making the Sierra Nevada more accessible. The Club began, in fact, to work against road and resort development, arguing that there was too much of it now.

Wilderness became a political issue. In the late 1940s Howard Zahniser, executive secretary of the Wilderness Society, helped to get a Congressional study of wilderness.[7] In 1951 he proposed a formal program of wilderness protection. The long-running fight over power development on the Colorado River, particularly the proposed dam at Echo Park, delayed matters, but in 1956 Senator Hubert Humphrey of Minnesota introduced the first wilderness bill. Nine years, a dozen hearings, and some fifty bills later Congress passed the Wilderness Act of 1964.[8]

Americans had long been fascinated by wilderness and were concerned that it was vanishing. Their demands for laws, though, also reflected their increasing use of wild land for recreation. Money, vacation time, cars, highways, and camping gear that made it comfortable to "rough it" drew them outdoors. The economic insecurity that had made them accept dirty air and stinking rivers as the price of progress was fading. An entire generation knew the Depression and its fears only as stories their parents told. They had jobs, knew they could get jobs, and expected prosperity to last. They could afford, and now demanded, environmental "amenities"—clean air and water and wilderness. When industrialization threatened the environment, they were ready to sacrifice a certain amount of economic development.

Concern that industrial development was wiping out wild animals and destroying wilderness was strongest among those already concerned with nature, but even dedicated indoor types found that the "environment" was affecting them in alarming ways. The first lessons came from atomic-bomb testing. Weapons tests had been conducted in the southwestern desert and on remote islands in the Pacific to keep them away from people. In 1950 scientists announced the discovery of radioactive isotopes from bomb tests in rain falling on New York State. Four years later several Japanese fishermen died from radiation poisoning after their ship was covered by fallout from a hydrogen bomb test.

Scientists were soon tracing fallout patterns in the atmosphere and the fate of the isotopes once they reached the ground. Their findings, presented in books and magazine articles and featured on the evening news, provided alarming lessons about connections in the environment. Winds high in the stratosphere carried radioactive isotopes from bomb tests around the world. When the poisons fell to earth they contaminated grass and trees. The animals that ate the vegetation concentrated the radioactivity. No place was safe. In the Arctic, caribou and reindeer grazing on the moss built up high levels of radiation. The Eskimos living on the animals had even higher concentrations. Closer to home, cows accumulated radioactive strontium-90 in their milk. American children, urged on by their mothers and the American Dairy Council, drank their quart of milk each day and built their bones and teeth with deadly isotopes. Bones were where blood cells were manufactured—and where leukemia struck. Fallout became a political issue. There were Congressional hearings, and in the presidential campaign of 1956 both parties promised to work for a test-ban treaty.

Then there was DDT.[9] When the "atomic bomb of insecticides" had replaced lead arsenate it had seemed all to the good. But, like the bomb, this scientific miracle soon proved to have unexpected and dangerous properties. It was more persistent than had been suspected, and food chains concentrated it. Related compounds caused cancer in laboratory animals, which raised further suspicions. The Food and Drug Administration's announcement, just before Thanksgiving in 1957, that it had found traces of a possibly carcinogenic pesticide on some cranberries caused public panic. Sales of cranberry sauce dropped and some grocery chains stopped selling the fruit. Officers of the Ocean Spray Cranberry Company ate handfuls of berries before press cameras and denounced the Food and Drug Administration. Vice President Richard Nixon had four highly publicized helpings of cranberry sauce at a dinner in Wisconsin Rapids, Wisconsin.

Things like this concentrated people's minds on ecological and environmental problems. At the same time, they were learning about wildlife via a flood of information. Books and magazines were important but movies and television added a new dimension. They showed animals; they did not describe them. They reached audiences, like young children, who could not yet read. They gave a more immediate experience of nature (or the illusion of one). Vivid, exciting, and readily available, animal films prepared Americans to "appreciate" nature in new ways.

Nature had been a popular subject for the movies since the days of hand-cranked cameras, but the products that appeared in the 1950s were qualitatively different. Color film, more sensitive films, improved slow-motion and time-lapse photography, and telephoto and macro lenses made it possible to see small creatures and large up close, to watch nocturnal animals hunting and feeding, and to follow sequences of action that took enormous patience to shoot but only interest to view. Studios like Walt Disney Productions took advantage of the new technology and their own resources to make a new kind of nature film. They could send teams of photographers off for a year at a time and then splice their efforts into films of a complexity, length, and realism denied individuals or small teams. Full-length nature films appeared that were more than incidents, often staged (a fight between a lion and a tiger), or pictures of large animals, or "safari" stories. They had something for everyone. There were exciting animal chases, time-lapse sequences of flowers bursting into bloom, sunsets and sunrises, migrating herds, and large, graceful animals posing nobly. There was even comic relief—ground squirrels chasing each other or the mating dance of the scorpions set to square dance music.

The films had a new perspective as well. They did not concentrate on individual lives but on the land and all its life. *The Vanishing Prairie* and *The Living Desert*, two of Disney's popular films, set the tone. They were often arranged as an exploration of an area during a year, unified by a narrator telling us how all the animals and plants lived together. The device encouraged people to think in terms of ecosystems (even if they did not yet have that name). The films also presented a new view of predation. Though often anthropomorphic, they did not, like *Bambi* and other cartoon features, show "cruel" predators and "innocent" prey. The black-footed ferret, for example, was featured in *The Vanishing Prairie*. He lived on the "cute" prairie dogs, but he was not a villain. Predation looked different in the new films. It was not cartoon violence with moral overtones. It was a careful stalk, suspense, a dramatic rush, and a cloud of dust. It was just another phenomenon of the exciting world of nature. Predators were part of the life of the land. Even when the narrator did not assure us that this killing was necessary, the presentation took place in a different setting than the conventional moral picture of the "cruel" wolf and the "innocent" deer. Nature movies reflected new scientific ideas and helped to shape public opinion.

Television was, if anything, more effective. Animal films had always been popular and the networks had many hours to fill. The result was a flood of shows that even the most suspicious parent would endorse. The genre has produced some of the most popular and durable shows on the air; in 1986 Marlon Perkins's "Wild Kingdom" celebrated its twenty-fifth year. Usually they did not teach ecology and they often concentrated on exciting events—the intrepid zoo director wrestling with an anaconda in the river—but they did familiarize people with animals and encourage an interest in wildlife. For a public with less and less opportunity to see animals, they provided a form of vicarious experience that shaped a generation's picture of the world around them.

Knowledge and concern grew together, but through the 1950s Americans made few connections among their environmental problems. It was not for lack of information. Vogt's *Road to Survival* had made the case that environmental problems all sprang from a single source. So had Aldo Leopold's *A Sand County Almanac*, which appeared just after the author's death in 1948. But while there were movements for wilderness and wildlife preservation, for curbing pollution and restoring the land, there was no common focus. That began to change in 1962, with the publication of Rachel Carson's *Silent Spring*.[10] Carson made environmental contamination a real and present danger. With the threat of nuclear war, she said, the "central problem of our age" is the contamination of man and the environment by pesticides. They disrupted ecosystems, killed animals, and could "even penetrate the germ cells to shatter or alter the very material of heredity upon which the shape of the future depends." We have introduced them into the environment and our food with "little or no advance investigation of their effect on soil, water, wildlife, and man himself. Future generations are unlikely to condone our lack of prudent concern for the integrity of the natural world that supports all life."[11]

The New Yorker magazine printed excerpts from the manuscript in June and July 1962—just as the thalidomide scandal became news. Pictures of hundreds of deformed babies, harmed by a medicine their mothers had taken, made Carson's charges about chemicals more credible and more vivid. The pesticide manufacturers' response to Carson's work only increased public interest. The National Agricultural Chemicals Association hinted darkly that Carson's ideas would help only the Communists, who would benefit from the destruction of American agriculture. Critics, some sponsored by NACA, said that Carson was impugning the integrity of workers in agriculture and industry. She was, they charged, a fanatic. If we followed her advice we would soon be back in caves living on roots and berries. Her ideas would destroy Western civilization as we know it. By the time the book appeared in the fall it was assured an audience.

Carson's popularity was not due entirely to sensational charges against persistent pesticides. In the same graceful prose that marked her nature essays, she identified a common cause for many of our problems and proposed a solution that went beyond bans on particular pesticides. We have, she said, treated nature as a set of independent pieces, each of which we might manipulate for our own ends. We have relied on our own scientific knowledge and technological expertise. Nature, though, is not inert, its parts are not independent. What we do reverberates through the "web of life." If we continued to act in ignorance and to disregard nature's laws, nature would react upon us. We would produce the "silent spring" of her first chapter, where no bird sang and a deathly silence covered the land. The remedy, though, was at hand. It was to abandon "brute-force" methods and forget about "conquering" nature. It was to take our place as citizens of a biological community, adapt ourselves to nature, and live by nature's laws. If we did, Carson promised, we would find, instead of a stark and silent spring, a harmonious world, full of life.

Critics who saw *Silent Spring* as a threat to Western civilization and Carson as a primitive calling for a return to the caves misread the book and misunderstood the author, but they did have cause for alarm.

Carson challenged concepts of "progress" and "the conquest of nature" that many people saw as the center of Western civilization and the hallmarks of American life. She did call for new ideas and new policies, for changes that would disrupt comfortable economic arrangements. Worse, people were listening. *Silent Spring* marked the beginning of the environmental movement. Carson provided the argument that connected the disparate causes; she helped fuse them together and to change the objective from particular things and areas to a policy that would preserve the ecosystems of the world. By the mid-1960s people were making the transition from education and thinking to action.

The Wolf and the Wilderness

There is a growing appreciation of an animal in an environment. Conservation is tending toward the *land* as the basic value. The animal is only a part of the ensemble. Many of us now feel that unless we awaken the public to the value of the forest itself, or the sage lands . . . and the sense of the *wildness*, we cannot hope to retain much of our wildlife.[12]

The public was awakening to the "sense of the wildness" and to the connections among all parts of that wildness. As it did, it found new value in wilderness and in the animal that had traditionally been the symbol of that land: the wolf.[13] It continued to be the symbol, but with a new twist. As the predator at the end of the food chain, it was a guarantee that the area had a "complete" food chain and was "undisturbed" wilderness. The wolf also changed. When the wilderness had been feared, the wolf had been fearsome. As the wilderness became wonderful, so did the wolf.[14]

Ethology, the science of animal behavior, developed rapidly in the interwar period and became a popular subject for nature magazines in the 1950s. People were accustomed to the idea that animals had a wide range of behavior and individual mannerisms. They were used, too, to the idea that people could establish links with animals. They applied this knowledge, and their new view of the wilderness, to interpreting and understanding wolves. Lois Crisler's *Arctic Wild* (1958) is a good example of this trend. During a year spent on the Arctic tundra photographing wildlife for a movie, she and her husband bought two wolf cubs from the Eskimos and took several others from a den to raise. Wolves, she found, "have gentle hearts but are not human-oriented." To know a wolf was to look into another world. She did not deny that it was a world of death, but it was also a beautiful, harmonious world, and the wolf was a part of it.[15]

Then there is the rehabilitation of the Custer wolf. This small, light-colored animal preyed on stock in and around Custer County, South Dakota, from about 1910 to 1920. The local ranchers credited it with extreme ferocity and an insatiable urge to kill and put a $500 reward on its head. When a Biological Survey trapper killed it in 1920, the news was "hailed with delight by stockmen throughout the region." W. B. Bell gave the event prominent notice in an article for the next issue of the *Yearbook of the Department of Agriculture*, "Hunting Down Stock Killers." There were no mourners.[16]

In 1966 Roger Caras published *The Custer Wolf*, the "idealized life story" of a family of wolves in South Dakota in 1910 and a reconstruction of the "outlaw's" career. He told the tale from the point of view of "this strange, tormented animal." He began with the wolf's world, where everything was "interconnected," where the death of prey animals is "the predators' gift to the land. Nature was in harmony, even with its thousand thousand deaths, and there was nothing savage, harmful, or tragic in all this." The savagery, harm, and tragedy all come from man, the wolf's one and eternal enemy: man, "eternally guilty of crimes beyond counting—man the killer, the slayer, the luster-for-blood—[who] has sought to expurgate himself of his sin and guilt by condemning the predatory animals."[17]

Man brings iron ore from the Upper Peninsula of Michigan and coal from Pennsylvania to make guns and traps. He imports strychnine from India. Man declares war on the wolf and destroys him. Caras ends with the man who killed the Custer wolf, with "[t]he photograph . . . old now, and a little yellow," and a few reflections on sentimentality, a "cardinal sin": "across the years since the blood dried and the hurt ceased to matter, I feel as if I have some slight understanding. . . . One has to be sentimental about a thing like that. At least I do."[18]

Farley Mowat's *Never Cry Wolf* (1963) shows an even more striking change in attitudes and greater development of another Romantic theme: the wolf as man's guide to nature.[19] The book is, ostensibly, a nonfiction account of the summer Mowat spent studying tundra wolves for the Canadian Wildlife Service (1948, though the year is never given). It is in fact a carefully constructed narrative of a spiritual experience.[20] When he went north, Mowat said, he had all the standard notions about wolves: they were bloodthirsty, savage, morose, cruel, and a danger to lone travelers. Over the course of the summer his ideas change. Watching "George," "Angeline," "Uncle Albert," and the cubs at their den and out hunting, he sees that they are gregarious, happy, and playful—not at all the "lone wolves" or vicious killers of legend. They do not terrorize their prey; they wander among the caribou, alarming the animals only when they come too close. Nor are they simply killers; they have a function in this world—culling the weak, the sick, and the stupid. They are part of a balanced and ordered world.

It is a world Mowat cannot enter. The psychological climax comes when he crawls into what he thinks is the empty wolf den and comes face to face with the mother wolf and a cub. Though the wolves do not threaten him—they cower away, in fact—all his fears return. He scrambles back out. Rage replaces fear. "If I had had my rifle I believe I might have reacted in brute fury and tried to kill both wolves." Fear passes. He is "appalled at the realization of how easily I had forgotten, and how readily I had denied, all that the summer sojourn with the wolves had taught me about them ... and about myself." Off to the east he hears a wolf howling, and it is "a voice which spoke of a lost world which once was ours before we chose the alien role; a world which I had glimpsed and almost entered ... only to be excluded, at the end, by my own self." A one-paragraph epilogue follows. In the winter of 1958–1959 the Canadian Wildlife Service placed bait stations and "coyote getters" in the area. Early spring thaws prevented the officer's return. "It is not known what results were obtained."[21]

Never Cry Wolf is a Romantic vision of man estranged from nature—but compare it with Seton's "Lobo, King of the Currumpaw." Both are popular tales, first-person narratives, of a man changed by his encounter with a wolf. Both stories begin with the author-observer pitted against the wolf. Seton is the expert "wolfer," with traps, guns, and poison; Mowat has his head stuffed full of old tales about ferocious wolves. Both end with the wolf guiding the man to a deeper experience of nature. The differences, though, are profound. Seton's encounter is a duel with an outlaw stock killer. Only one outcome is possible; the wolf has to die ("It cannot be otherwise," Seton intones).[22] Mowat meets the wolves in the context of a scientific study. This is not man against wolf. George and Angeline are not stock killers. Their

(presumed) deaths are not necessary. They die because of a bureaucratic decision to kill wolves to satisfy hunters who see the animals as
competitors. Their deaths are not individual, but the result of an im-
personal administrative program sowing the land with poison. Seton
feels nostalgia for a passing world and sorrow at the death of an animal lured to destruction because it was faithful to its mate. Mowat is
overwhelmed by a sense of his own failure to reach the moral level of
the wolves and to enter their world. Beneath that is anger and con-
tempt for those who destroy nature.

The public has responded enthusiastically to the message of the
good wolf. In the twenty years since its publication, *Never Cry Wolf* has
sold a million copies. In 1983 Walt Disney productions made it into a
movie. Barry Holstun Lopez assembled scientific reports, Indian lore,
historical studies, and personal observations to argue that men have
created wolves in the image of their worst fears.[23] R. D. Lawrence has
produced some dozen popular books on wolves, the latest entitled simply *In Praise of Wolves*.[24] The wolf, once pariah, has become paragon.

Wolves have become tourist attractions, though the chances of seeing
one are slight. When Canadian biologist Douglas Pimlott began giving
lectures on wolves in Algonquin Park, Ontario, audiences were "standing room only." In August 1963, the park newsletter invited people to
attend an "evening of wolf listening." The invitation produced a traffic
jam of 180 vehicles and some six hundred people—to the astonish-
ment of park staff. Since then "howling sessions" at Algonquin have
been popular features. Bands of tourists howl like wolves in the hope
of answers from the wild (the sessions remain popular because wolves
will respond even to poor imitations of howling).[25]

Ideas into Action

"In no country in the world has the principle of association been more
successfully used or applied to a greater multitude of objects than in
America."[26] Thus Alexis de Tocqueville described the tendency of
Americans in the early nineteenth century to use voluntary societies to
express their will. Their descendants continued the tradition, and interest in nature has spawned a bewildering assortment of organizations. One consequence of Americans' rising interest in nature has
been an increasing number of groups devoted to saving it. In 1945 the
Fish and Wildlife Service's listing of nongovernmental wildlife groups

had fifty-six names, most of them of well-established organizations. Numbers rose slowly through the 1950s, then accelerated. The NWF's *Conservation Directory*, which supplanted the Fish and Wildlife Service's list, swelled to over three hundred by the mid-1970s and almost four hundred a decade later,[27] including quite specialized organizations. Before World War II Ducks Unlimited and the Bison Society had been almost alone in their concentration on a single species or kind of animal. By the 1960s there were many. Some, like BASS, the Bass Anglers' Sportsmen's Society (organized in 1968), or The Foundation for North American Wild Sheep (1977), were concerned with game species. Others, such as the North American Bluebird Association (1978), favored songbirds. Still others took up the cause of less familiar forms. There was the Desert Tortoise Preservation Committee (1974) and the Xerces Society, interested in arthropods (1971). In the 1960s several were formed to protect the formerly despised "vermin": the Society of the Preservation of Birds of Prey (1966) or the Wild Canid Survival and Research Center/Wolf Sanctuary Association (1978).[28]

Many more people joined these groups. Several of the most popular doubled during the height of environmental enthusiasm. More impressive, if less obvious, was the general increase.[29] Larger memberships produced changes within the organizations. Magazines expanded and acquired slick paper, more and better photographs, and more sophisticated layouts. Reports to the members now spoke of staff offices in Washington, bills before Congress, and the need to write one's congressman or senator. Wildlife organizations educated and informed. They provided a sense of community and shared values. They also, despite the judgments of tax lawyers and the protestations of their directors, lobbied for what they considered desirable wildlife policy. It might be as little as the executive director lunching with the assistant undersecretary, but it was effective. Quite aside from the expertise they were acquiring, wildlife organizations now had large numbers, and their members were the kind of people who could make a difference. They were overwhelmingly white, economically well off, and well educated. They spoke, and they expected to be heard.

By 1960 wildlife advocates were bringing specific demands to Congress, where laws were made, and the Fish and Wildlife Service, where laws became policy. One major complaint was the predator-control program. Its goal, a western pasture free of "varmints," was at odds with the new vision of man and nature in harmony. So was the favored method of control, poisoning. It killed nontarget species, some of them already endangered, and disrupted ecosystems. Another important concern was protection for endangered species. Extinction was a dramatic and irreversible change in the biological world. The rate at which species were vanishing was increasing. What could we do to save our (now precious) biological heritage? Both goal and method brought the PARC under attack, and opponents were not protesting "excesses." They wanted poisoning stopped altogether and the policy of "control" that saw coyotes as "varmints" reversed. The primary villain was the new poison the Fish and Wildlife Service had introduced after World War II—Compound 1080.

Interlude

Values and Varmints, II

At first glance the fight against compound 1080 looks much like the battle the mammalogists had waged against PARC operations in the 1920s. There were even some of the same people—for example, E. Raymond Hall came before Congress again to protest poisoning. There were, however, also profound differences. The public was involved, as it had not been in the earlier fight. Science was more central; ecology was the rallying cry of poisoning's opponents. The coyote, too, had a new place. Protests now were not just that "innocent animals" were being killed. People explicitly valued the "coyote itself." The "environmental" sentiment behind the new anti-poisoning campaign would change not only the PARC but the Fish and Wildlife Service of which it was a part.

CHAPTER EIGHT

Poisons and Policy
New Values for Varmints,
1939-1964

The public's new standards put the PARC's operations and goals in a new light. Exterminating prairie-dog towns with strychnine-treated grain, poisoning coyotes wholesale, clubbing the pups to death, and shooting wolves had been acceptable, even laudable, when Americans were conquering the wilderness and making the land productive. When they wanted to preserve wilderness and wildlife, when they were concerned about the suffering of animals, these practices seemed cruel and destructive. To make matters worse, the PARC had ex- panded its operations after World War II and introduced a new poi- son, sodium fluoroacetate (usually called Compound 1080 after its wartime laboratory test number). It was extremely toxic to rodents and canids. The normal lethal dose for coyotes, for example, was a few mouthfuls of horse meat treated at the rate of 1.6 *grams* of poison per hundred *pounds* of meat. In the late 1940s the PARC began using the chemical in large quantities. By the early 1960s critics would be charging that the chemical and the control program threatened the ecosystems of the American West and endangered many species in the region.[1]

The fight over 1080 and PARC policy, which began in the late 1950s, was at bottom a matter of values. On the one side were the woolgrowers, committed to making the land productive. They saw 1080 as part of the technology that sustained men in the face of a hostile nature. They believed that the PARC program was an essential government service. Environmentalists regarded the chemical and the control program as examples of our reckless disregard for the natural world on which we depended. Advocates of humane treatment con- demned the work as government-subsidized cruelty. To the ranchers a ban on 1080 and other poisons would be a betrayal of the western stock industry and of American values. To the other side a ban seemed the first step toward a sound wildlife policy. It was a declaration of our commitment to an environmental ethic and to humane treatment for wildlife.[2]

There were few side issues. Ten-eighty was quite poisonous to humans, but the baits contained such a low concentration that they would not be lethal (in any case, it is hard to imagine even a starving person eating a two-months-dead horse). There were none of the worries about industrial exposure, cancer, or birth defects that made DDT, PCBS, vinyl chloride, or atomic radiation as much human health and safety issues as environmental problems. Nor was this a case of newly discovered drawbacks calling into question accepted practices. Ten-eighty's effects on wildlife were known—if not fully appreciated—when the Fish and Wildlife Service began using it.[3] Opposition arose because Americans' values had changed. Practices people had ignored, condoned, or even approved they now saw as inhumane or destructive.

The Push for Poison

The same conditions that had generated the PARC program early in the century and had made poison attractive enlarged the program after World War II and made 1080 its mainstay. Ranchers had always counted on subsidies. They had relied on free or low-cost grazing on federal land, tariff protection, and help in killing predators. This last was particularly important and sensitive. Losses to coyotes, bears, and mountain lions—a peculiarly personal source of loss—were visible. Stockowners saw predators as enemies, and they believed in fighting them to the death.

That the West was large and funds were limited had from the first encouraged the Survey to use poison and to concentrate research on more effective ways to use it. The Depression increased that pressure. Ranchers wanted predator control, for it allowed them to cut the amount they spent on herders, fences, and lambing sheds. The PARC "sowed" isolated areas with baits dropped from airplanes. It adopted the "coyote getter," a device that fired a charge of sodium cyanide into an animal's mouth when it picked up a bait.[4]

In 1937 the Survey's Denver Wildlife Research Laboratory began testing a new way to protect sheep on lambing grounds high in the mountains. Because snow closed the passes until shortly before lambing began, it was impractical to kill the coyotes in the spring. Researchers decided to put the stations in before the snow fell and let them kill coyotes all winter. Several years of tests showed that the scheme worked, but the best poison turned out to be thallium sulfate. After the controversy in California in 1931 Harold Anthony had written to Grinnell that "in Washington they consider thallium so much dynamite and are afraid to use it."[5] Fewer people would see its effects in Wyoming mountain pastures than in California fields, but the stations would still cause controversy. In 1944 Weldon B. Robinson, in charge of the project, reported finding dead around the stations (presumably poisoned) 673 coyotes, twenty-four dogs, three badgers, eight ground squirrels, four weasels, one cat, four eagles, four hawks, and nine magpies. The counts were, he admitted, partial and biased toward the coyote column.[6]

There was, though, tremendous pressure on the PARC to use thallium bait stations. World War II had increased the demand for wool and mutton even as it made many standard control strategies more difficult or expensive. The draft and high wages in defense industries pulled sheepherders off the range. War in the Pacific cut off supplies of strychnine from India. Steel rationing meant fewer traps, military demands meant less rifle ammunition. The bait stations were also working; Robinson reported that they had cut losses to predators in the test areas by 75 to 96 percent. Ranchers, he added, were "very insistent in their demands that the use of the poison be continued." The results "have been so convincing . . . that to withdraw . . . [it] . . . would result in serious repercussions in the control program." If the PARC did not go ahead, he warned, the sheepmen would. That would mean widespread misuse, and dead hunting dogs and ranch pets would reflect badly on the agency—regardless of its involvement. The Service, he concluded, had to use thallium. It also had to impose extraordinary restrictions to reduce the public reaction.[7]

Keeping the lid on was difficult. A year after Robinson made his recommendations, his superior, E. R. Kalmbach, sent a memo to Washington. Robinson wanted to publish his research results. Publication, Kalmbach argued, would imply official approval of the procedures, invite criticism, and lead to increased use. He was pessimistic about the agency's ability to control the chemical. "The gap between approved and announced procedures and actual field practice, influenced by pressure from the outside, is destined to increase," he said—

a delicate allusion to the ranchers' tendency to press for more poison and for field agents to cut corners on safety.[8]

The agency was concerned, it should be noted, not about opposition to poisoning, still less to predator control. It was worried that cruel methods of killing and the destruction of nontarget wildlife would bring a reaction from humane societies and wildlife lovers. They could not change policy, but they could force the PARC to spend time and money defending itself. Still, publicity in one form or another was unavoidable. The Fish and Wildlife Service suppressed Robinson's research results, but there would be other papers, scientific or popular. It looked as if thallium could not be safely used and could not be banned.

New Poison, Old Program

Even as these problems arose, compound 1080 promised to solve them. Wartime tests for rodenticides (needed to control rodent vectors of typhus and plague) showed that it was much more toxic to rodents and canids than to other forms of life. The PARC began field tests in 1945. Results were so good that at least one PARC agent thought it might mean the end of all predators.[9] Ten-eighty was not a perfect poison, though, by any means. The pure compound was deadly to humans and there was no known antidote. It was water-soluble, and so might contaminate streams, wells, or rivers. It was also extremely stable. The first field tests, done in Kern County, California, showed that animals eating poisoned ground squirrels and prairie dogs could die from unassimilated poison in the rodents' system or grain remaining in their cheek pouches. In a memo on 1080 use, PARC chief Dorr D. Green said that "Careful handling and the danger of secondary poisoning cannot be overemphasized" (emphasis in original). This was particularly true "in the case of canids that are feeding on the rodent victims of 1080. This is largely due to [their] high susceptibility." The "canids" the PARC was concerned about were farm dogs.[10]

Other species were less sensitive, and the PARC thought this would protect them. Vultures could tolerate much higher doses than coyotes and they regurgitated when "in trouble." On this basis the PARC considered that bait stations posed no danger to the California condor. (For obvious reasons, no one tried poisoning a California condor. The PARC extrapolated from work on a related species, the Andean condor.) Kalmbach noted that a condor had been seen eating poisoned bait in Kern County with no observed ill effects. He was less sanguine about eagles and hawks, but said there were indications that "their relatively greater resistance to 1080 [compared to canids] is giving them a measure of protection when feeding on poisoned rodents."[11]

Though 1080 seemed ideal for overwintering bait stations, the potential dangers loomed large enough to delay approval until 1947. When it was approved it was subject to a stringent set of regulations. The Fish and Wildlife Service negotiated a "gentleman's agreement" with the manufacturer, Monsanto Chemical Corporation. The company pledged to sell 1080 only to the Fish and Wildlife Service, other official agencies engaged in pest control (primarily western states and counties), and qualified exterminators. For predator control the Service decided to use 1080 only west of the 100th meridian—the less-settled and drier part of the country—and only where predation was a major problem and other methods were not working. It would not use 1080 in aerial baits or drop baits—only in high mountain, winter bait stations. These were to be closely controlled. Coyotes ranged widely, so stations would be limited to one per congressional township (thirty-six square miles). Baits were to be located away from roads, improvements, and water, and the regional director would approve each location. Agents trained and authorized to use 1080 would personally put the baits in as late in the fall as possible, remove them in the spring as soon as the snow melted, and burn all remains.[12]

Restrictions proved to be more impressive in theory than in practice. The woolgrowers' demands very quickly pushed the Service toward large-scale use of 1080. In 1949 Kalmbach complained from Denver that the initial guidelines had been discarded. The Service was "promoting the use of 1080 far beyond the limits that have been recommended through adequate research." He was particularly upset about references to tests and experiments. These gave the impression that work done at the Denver laboratory justified the new look. This, he said, was not the case. He went on to diagnose the trouble: "It is not surprising that my viewpoint is quite consistently at variance with that of those engaged in the operational program. In actual practice, their primary responsibilities are to meet the wishes of a single industry." But the Service was making a mistake: "Whereas it may seem logical to heed the opinions of those interests from which much of our finances emanate, it is my conviction that we have gone too far in that direction. Already the unorganized opposition is being heard, and I feel that in time this will take more coherent shape."[13]

In another memo, two years later, he showed that the number of stations (15,289 in 1949 and 16,668 in 1950) and the range each covered meant that about half the West had a station in reach. Some states were almost saturated—91 percent of the range land of Idaho, 83 percent of Utah, and 71 percent of Nevada were covered. All this, he pointed out, had taken place without reducing the use of traps, guns, and strychnine, and a scant four years after the Denver lab had recommended and the Fish and Wildlife Service had approved the use of 1080 "primarily on acute predation areas where other methods have not gained the desired degree of control."[14]

Kalmbach was the most quantitative critic, but not the only one. Clarence Cottam, head of the Division of Wildlife Research, fought for more tests before 1080 was approved. Once it was in the field, he pressed for stronger controls on its use. In 1951 he complained of pressure on the Service to push 1080. The agency, he said, was acting on the principle that the only good coyote was a dead coyote: "I am fearful that such an extreme attitude is going to get the Service in trouble."[15] He was worried, too, about the poisoning of nontarget wildlife.

The files on violations bore out his concern. In 1949, on the margin of a letter about these matters, an agent wrote: "As for the sheepmen moving our 1080 stations, this happens more or less all over, and I do wish we could do something to stop them from doing it." In another case a rancher told a field agent he would destroy three stations on his property (a violation of procedures that would save the agent a trip), but he failed to do the job. Some agents, it appeared, were deliberately leaving the stations out all summer.[16] The "gentlemen's agreement" proved ineffective. County boards could still purchase 1080, and they were not always scrupulous. In 1953 officials in Campbell County, Wyoming, purchased two and a half pounds of the chemical for local predator control. This was enough, PARC agents later estimated, for the entire state. A decade later a memo noted that federal use of 1080 in California had been 14.7 ounces, and state use (mainly for forest and rangeland rodent control) had been 6,000 ounces. There must, the memo writer added, be "much irresponsible use here."[17]

Cottam blamed officials in Washington. They were, he said, willing to bend or break regulations for the sake of the ranchers and field men and make policy on the basis of public relations. "I well remember," he wrote Starker Leopold in 1956, "instances where there were flagrant violations of Service policy by field personnel of the Service, and I believe in every case [C. C.] Presnall and [Dorr] Green were not only sympathetic with these violations but gave support to the field men who violated the regulations."[18] The problem, though, went beyond individual attitudes. It was structural. Any agency would find it difficult to make policy without regard to "the opinions of those interests from which much of our finances emanate," and the woolgrowers contributed more than money. They provided political support. They gave PARC agents permission to work on their land. Field agents came from the same area as the ranchers and shared their values. Even if they did not, they had to live with the ranchers. The woolgrowers generally got what they wanted.

And they wanted poison. They had to cut their costs or cut their flocks. Most chose more 1080 and fewer herders.[19] With the PARC's cooperation it worked. Doing research for an article on 1080, a PARC staffer found a rancher who had raised "7,000 lambs that spring without a single known loss to predators! And this was accomplished in lambing on the open range, no herders, no fences!" When he asked the man what would have happened if he had tried that twenty years ago, he "just grinned and shook his head." Pressed, he said that if he "had been fool enough to try lambing without herders and dogs in that area in 1946—when the 1080 program began—the coyotes would have eaten him out of house and home."[20]

The Service ignored inside critics and relied on the same tactics the Survey had used in the 1920s to defuse outside protestors. When mountain-lion hunters complained that their dogs were finding bait stations—with fatal results—the Service expressed its regrets, pointed out the need for poisoning, and promised better publicity to warn hunters what areas to avoid. In the late 1940s there was a spate of anti-1080 articles in sportsmen's magazines, claiming that the new chemical was killing off wildlife. The sport of thousands was being sacrificed for the benefit of a few ranchers. The Service asked for space in each publication to refute these arguments. Ten-eighty, it said, was selective and fast-acting. Poisoning was necessary for the livestock industry and it helped game populations. Guidelines for the program and regulations on the sale, distribution, and use of 1080 kept the public and nontarget wildlife safe. When scientific associations became restive, as the American Society of Mammalogists did in 1950, the Service sent a representative (usually Clarence Cottam) to discuss the matter.[21]

By the late 1950s the "unorganized opposition" Kalmbach had predicted was making itself felt. The usually cooperative Forest Service had begun to balk at predator poisoning on its lands. A note of bitterness crept into the field agents' letters to Washington, with complaints about "backsliding" and "duty-struck" foresters. One district agent sensed a conspiracy. The rangers, he wrote, were "being pushed along very much by a few individuals in the Forest Service whose activities are well known to me."[22] Another advocate of poisoning dismissed critics as people "who are opposed to any killing for religious or other reasons, those who want to protect predators in certain desert areas, hound dog associations that have an unwarranted fear of the product, or . . . those who do not understand the problem."[23]

At the time this was an accurate list. Opposition was growing but opponents were scattered. The hound-dog owners did not object to poisoning, even vigorous poisoning. They just wanted to keep their animals safe. Others, echoing the protestors of thirty years earlier, wanted controls on the program. A larger group, though, wanted to end poisoning altogether. Some were "opposed to any killing for religious or other reasons." Others were concerned that poisoning was disrupting the ecosystems of the West. But none of these seriously threatened the PARC. Hound-dog owners were a small minority, and so were humane advocates. Unless poisoning's opponents could find an issue that would rouse the general public, they could do little, and there did not seem to be such an issue at hand. The argument that the program was disrupting ecosystems was the strongest, but few scientists were willing to say, on the evidence available, that the PARC program, as it was being conducted, was so harmful that it had to be stopped.

Endangered Species

Ten-eighty's opponents found their issue in endangered-species protection. Extinction was a dramatic example of the destruction of nature. It had been a distant danger. When people thought of extinction they thought of the past, of the passenger pigeon and the dodo. As habitat destruction and pollution increased, so did the rate at which species vanished. As people learned about the problem, often from

119

the wildlife conservation groups they had joined, they became concerned. The whooping crane had been in trouble since the early part of the century, and since 1948 the United States and Canada had been working to save it. Interest rose dramatically in the 1950s. By the middle of the decade the Fish and Wildlife Service was holding press conferences, and newspapers and television reported the annual count like a baseball score.[24] Ecology added a new argument for protecting these species. If diversity meant stability, if each species had a role and a function in the ecosystem, then the loss of any was a potential disaster. Humane societies joined the coalition. Extinction, they said, was the ultimate cruelty toward wildlife. The issue also attracted the public because, it seemed, something could be done. We had saved the buffalo, and its herds were still increasing. We had saved the herons and egrets; ducks were recovering; and the whooping crane was holding on. Why could we not help other animals and birds?

The PARC's first problems with this new view of wildlife arose over an animal that had never attracted much notice, from it or from the public—the black-footed ferret. The black-foot is the largest weasel on the continent (up to two feet in length) and it is quite distinctive (dun-colored on the back, shading to white on the belly and marked with black "boots" and a "mask" over its eyes). It is, however, nocturnal, shy, and, apparently, has never been very abundant—though scientists found a colony of almost one hundred in Wyoming in the early 1980s.[25] Audubon described the species in 1851 on the basis of a single skin that was later destroyed in a fire. For a quarter of a century there were no other sightings or records. A few people suggested that the animal was a hoax. Only in the late 1870s, when Dr. Elliott Coues advertised for specimens in the *American Sportsman*, did more come to light.[26] Even Coues found very few. He reported that he had spent the summer of 1876 doing a natural-history survey "through the region supposed to be its centre of abundance, where [two specimens had been] secured. . . . [But] I failed to obtain a sight of it, though I was in the midst of prairie-dog towns, and continually on the watch for this particular animal."[27]

Ernest Thompson Seton's monumental study of North American mammals contained little about the ferret. Its range, Seton said, "coincides nearly—maybe exactly—with that of the Prairie dogs." Of its habits, "[v]ery little is on record concerning the habits of the Black-foot. We know that it lives as a parasite in the dog-towns, lives like a

Mouse in a cheese, for the hapless Prairie-dogs are its favorite food, and its den is one originally made by the Prairie-dog for himself." For the rest, he confessed ignorance. Forty years later Stanley Paul Young could say little more. Even physical evidence was scarce. The combined collections of the Fish and Wildlife Service and the Smithsonian Institution, "probably the most representative collection in the entire world," had only sixty specimens. In 1953 Young said that a specimen just received from South Dakota would "give us the first full and complete skeleton of the animal, and also our first chance to examine the stomach of the black-footed ferret."[28]

Even in 1900 the ferret's future had been dim. "The Blackfoot," Seton said, "always rare, is becoming rarer. Now that the big Demon of Commerce has declared war on the Prairie-dog . . . the Ferret, too, will pass. Is this a pleasant thought or a sad one? Does it inspire you to some action? If so, what? There is but a little time left in which to do it."[29] No one, though, did anything. Not until 1953, when the American Committee for International Wildlife Protection asked the Fish and Wildlife Service about the status of the animal, is there even a hint in the files that anyone was thinking about it. The next year, though, PARC chief Dorr Green told John Farley, chief of the Fish and Wildlife Service, that his people had "for some time" been thinking of methods to preserve this scarce animal: "It is felt that active work by the Service in preservation of the black-footed ferret would have a decidedly favorable public relations value and would be timely in view of current correspondence and publications on the subject outside the Service." Two months later Farley wrote to the regional directors that "the Service would take active and vigorous steps to preserve this unusual species."[30]

The steps were not too active. The Service planned to take some specimens in live traps, but, as an agent explained to officials in South Dakota, this would take some time: "We do not plan to make any special effort to take them except as we might run across them in the course of normal prairie dog control operations."[31] Some were trapped and did a stint before the cameras for Disney's *The Vanishing Prairie*. They were then released in Wind Cave National Park. One was run over by a car and the other disappeared. A note attached to a secondhand report of the incident urged an end to trapping because of "South Dakota's feelings and the Disney–Wind Cave fiasco."[32]

In 1963 Ian McTaggart Cowan delivered a paper on endangered

species at the International Congress of Zoology. He said that there had been no sightings of the ferret since 1955. The Service's only response was to send a memo to the regional directors asking for information on recent sightings. It explained the lack of information by saying that "we have been reticent [*sic*] to disclose the whereabouts of ferrets and have discouraged their capture, even for scientific purposes—all with the intent of aiding their preservation."³³ There was, though, no program, no plans, and no research. All that was preserved was the poisoning of prairie dogs.

By the early 1960s public pressure for endangered-species protection pushed, or allowed, the Fish and Wildlife Service to take more active measures. In 1962 it set up an endangered-species committee. Two years later it made this an Office of Endangered Species, charged with the task of preparing a domestic equivalent of the IUCN's "Red Book" of endangered species.³⁴ The ferret was among the Office's first cases, and it was already attracting some publicity. When the Service began a survey of the existing ferret population in 1965, the Assistant Secretary of the Interior for Fish, Wildlife, and Parks, not the head of the Fish and Wildlife Service, made the announcement. The same year the Service decided that "[i]n view of the current keen interest in the preservation and protection of rare and endangered species, it is apparent that we must expedite our search for less . . . hazardous chemicals than Compound 1080."³⁵

The "dual responsibility for providing necessary safeguards for rare and endangered species, of which the black-footed ferret is one, and also to provide proper supervision and guidance of programs of animal control" made that search urgent.³⁶ The Sioux on the Pine Ridge reservation in South Dakota wanted assistance in poisoning prairie dogs on their land. However, one-third of the recent ferret sightings in the country had come from the reservation. The Fish and Wildlife Service, caught between the Indians and the wildlife advocates, had simultaneously to carry out a research program, continue control operations, find ferrets (if they were there), and avoid unfavorable publicity. It had to negotiate with the Bureau of Indian Affairs and deal with the Defense Department on a proposed land swap to set up a ferret sanctuary. It had to coordinate its plans with the South Dakota Fish and Game Commission, which controlled resident wildlife in the state and placated scientists and wildlife lovers.³⁷

THE FERRET CASE was one example of what was rapidly becoming a common kind of conflict. When wildlife conservation had meant saving a few species (and those in parks) or encouraging the preservation of songbirds and farm game, it had been possible for the Fish and Wildlife Service to protect wildlife and poison "varmints." When people demanded that all species be preserved, when they wanted a full complement of wildlife as part of the landscape, the contradictions among the Service's missions became too obvious to ignore. It is rare, though, that government agencies perish from such contradictions, particularly when the people within and without have the same values. Fish and Wildlife officers were better informed than the public about endangered species, habitat destruction, and the general loss of wildlife. Most of them liked wildlife and were anxious to save it. Many of them shared the public's distaste for the PARC's program. Even some who worked in the division were skeptical of the efficacy of broadcast poisoning and worried about its effects on other wildlife. When the public sought to put its new values into practice, when it began to press for a wildlife policy that would preserve ecosystems and save all species, the Fish and Wildlife Service was not only a target of attack; it was also an ally in the reshaping of its responsibilities and programs.

PART III

New Ideas, New Action

An appreciation of the place of predators in the eco-system, a determination to preserve the natural environment, and revulsion at animals' suffering did not of themselves end the poisoning of coyotes or lead to the protection of wolves. Wildlife's defenders had to translate their concern and public support into political pressure. They had to get laws passed and executive orders issued. Then they had to see to it that the agencies responsible for action actually took it. This is the story of the last twenty years—and of the present. We are still trying to understand this new way of looking at nature, still learning to "interpret nature aright" and to do right by it.

CHAPTER NINE

Ending the Poisoning

1963-1972

Only the woolgrowers stood against the growing public opposition to poisoning, but they were a formidable force. Predator control was, as they conceived it, a matter of economic life or death, and they held important strategic ground. Agriculture and Interior Department policy had always been important to Westerners, who were well represented on the committees that reviewed these agencies and determined their budgets. Even if they were a minority they could, unless the issue was of national importance, block the legislative process until the majority, composed of people who had other issues they wanted passed, gave in. Wildlife protection groups realized this and avoided a direct confrontation. They went instead to the Department of the Interior. They were part of the department's complement of political allies and could expect at least a sympathetic hearing.

Secretary of the Interior Stewart Udall was sympathetic, but he could hardly act on the basis of a few complaints. He did what any bureaucrat would under the circumstances: he called for a study. He asked his Special Advisory Board on Wildlife Management to review the predator and rodent control program. This might have been a political move; one of the most important functions of committees is, after all, to bury things, but Udall was looking for an answer. The group was well equipped to give him one. Usually called the Leopold committee after its chairman, Starker Leopold (Aldo's eldest son), it had been formed in 1962 to advise the Secretary on management of the elk herd in Yellowstone National Park. Its members had varied backgrounds. Starker Leopold had graduated from the University of Wisconsin in agriculture, spent a year at the Yale School of Forestry, and gone on to do his Ph.D. work at the Museum of Vertebrate Zoology in Berkeley, working with Joseph Grinnell. He had, by 1962, twenty years of experience on projects ranging from managing turkeys in the Ozarks to investigating Alaskan game for the Conservation

Foundation. Stanley Cain had a Ph.D. degree in plant ecology from the University of Chicago. Thomas Kimball had a B.S. degree in soil conservation and had been coordinator of federal wildlife work in Arizona for the state fish and game department, then director of the agency, then head of the Colorado Game and Fish Department. He was in 1962 executive vice president of the National Wildlife Federation.

There were two government biologists on the committee, both "old-timers." Clarence Cottam had joined the Bureau of Biological Survey in 1929 and risen to head the Division of Wildlife Research. He had been involved with Compound 1080 since the wartime tests. He was a loyal agency man, but also a strong inside critic of poisoning. Ira Gabrielson had changed with the times. He had joined the Biological Survey in 1915 as a "gopher choker" and had backed the poisoning program, resigning from the American Society of Mammalogists in 1931 in protest against the Society's condemnation of the Survey.[1] As chief of the Survey from 1935 to 1940 and head of the Fish and Wildlife Service for six years after that, he had taken a different tack, writing popular articles that stressed the agency's wildlife-conservation work. One detailed the Service's efforts to preserve endangered species. His *Wildlife Conservation* (1941) incorporated the latest information on predator-prey relationships. By the mid-1940s he was defending predators, citing Adolph Murie's work as evidence that the animals had a function and should be saved.[2] In 1946, when he left the Survey, he became head of the Wildlife Management Institute. The year before he joined the Leopold committee, he helped organize, and became head of, the World Wildlife Fund.

Times had changed in other ways. In 1924 Secretary of Agriculture Henry C. Wallace had turned to representatives of groups using the Kaibab plateau to advise him about the Kaibab deer; in 1963 Secretary Udall used a group of scientific experts. They were now the accepted arbiters of policy. In the 1920s the Survey had strongly supported poisoning. Now even those who, like Gabrielson, had been behind the policy were against it. Predator control policy had been made by a few interested agencies and groups; there had been no public input or outside scrutiny. Now there was an outside group charged with assessing the program and advising the Secretary of the Interior what changes should be made. The woolgrowers still had influence. They no longer had undisputed power.

The committee presented its report in March 1964, to Secretary Udall and to the annual North American Wildlife Conference. It provided ammunition for a full-scale assault on the poisoning program. The Division Predator and Rodent Control had defended its work as necessary and scientifically sound; the committee found that it was neither. There was little evidence, it said, to show that poisoning actually protected livestock. What there was suggested that the program was not uniformly effective and cost more than it was worth. There was even less support for other justifications. The danger from animal rabies was "negligible." The argument that native wildlife needed protection from predators was "supported weakly, if at all" by the last twenty years of research on wildlife. Control was not based, the committee found, on any demonstrated need to curb predator populations, but "on the subjective judgment of the PARC field men or supervisors in conference with livestock operators and agricultural officials." Some supervisors were "careful and conscientious," but "we have abundant evidence that others willingly support almost any control proposal in which someone is interested enough to contribute matching funds." The program has become "an end in itself and no longer is a balanced component of an overall scheme of wildlife husbandry and management." The PARC was a "semi-autonomous bureaucracy whose function in many localities bears scant relationship to real need and less still to scientific management."[3]

The committee called on the PARC to reassess its goals in the light of new public attitudes toward wildlife, use the minimum of control consistent with economic interests, supervise field operations more closely, and resist building the program for its own sake. So far, the suggestions were much like those the mammalogists had made in the 1920s. Leopold's committee went farther. It recommended the appointment of a permanent advisory board to represent all interests and make the Secretary "aware of sensitive problems and divergent interests."[4]

Udall accepted many of the recommendations. He appointed a new director, John Gottschalk, for the Fish and Wildlife Service. He found a new head, Jack Berryman, for the PARC. They replaced, Berryman recalled, thirty-two of thirty-seven state supervisors and all but two of the regional administrators. They set up a training program for field agents to acquaint them with the new policy and the reasons behind it. They began cracking down on careless use of poison (Berryman

remembered with particular horror the poisoning of two dogs belong-
ing to the Colorado representative of the Defenders of Wildlife).[5] The
PARC became the Division of Wildlife Services and adopted a new
policy statement, "Man and Wildlife," that incorporated much of the
Leopold report. Predator control, the document said, should aim at
removing "the offending individual animal wherever and whenever
possible"—an explicit rejection of the policy that had guided the
PARC for the last forty years. Bears and mountain lions were "particularly prized parts of the Nation's wildlife heritage" and would be
killed only when there was a "documented reason."[6] Predator control,
it seemed, was off to a new start.

The crux of the matter, though, was what would happen in the
field. Since the 1920s critics had claimed that government hunters
would do as they pleased, regardless of official policy. The bureau's
reform plans, therefore, met with some skepticism. Starker Leopold
thought that the final draft of "Man and Wildlife" was "all right" but
that the real test would be in its administration. Thomas Kimball, writing to Gottschalk, was less formal: "Knowing some of your co-workers
in the West who will be required to carry out these policy changes, and
of their political acumen, . . . [watching you work] . . . should prove a
most interesting spectator sport. I don't envy you your job."[7]

The report was the first in a series of setbacks for the woolgrowers.
Another came in the spring of 1966, when Representative John Dingell of Michigan, a member of the House Committee on Merchant
Marine and Fisheries, opened hearings on predator control policy before his Subcommittee on Fisheries and Wildlife Conservation.[8] Dingell's purpose was, ostensibly, to help Congress find out what initia-
tives were "being taken by the Federal Government to properly
manage all wildlife resources." Dingell was in fact using the hearings
to rally opponents of poisoning and put pressure on the Fish and
Wildlife Service.

Woolgrowers and western senators and representatives came before
the subcommittee to plead for help. The "greatest crisis facing the
sheep industry in Wyoming today," said Representative Teno Ronca-
lio, "is the lack of adequate control of predators." Representative
O. Clark Fisher of Texas spoke of the "serious menace" posed by the "vicious and relentless Mexican [golden] eagle." Wildlife defenders were
equally fervent. E. Raymond Hall, now head of the Museum of Natural History at the University of Kansas, condemned poisoning as vig-

orously as he had thirty years earlier. It was not necessary, he said, and pointed to the success Kansas and Missouri had had in training farmers to trap individual animals. Poison did destroy wildlife and disturb the balance of nature (that term again!). Alfred Etter, Colorado representative of the Defenders of Wildlife (his were the poisoned dogs), used official maps to show that field agents were constantly violating regulations.[9]

The charges and countercharges were similar to those raised in the 1920s, but the political climate was different. When Hall and Howell had testified against the Survey's ambitious plans in 1930, congressmen had generally been unsympathetic and the public indifferent. This time Dingell opened the hearings by referring to "well known excesses" and "indiscriminate trapping, shooting, and poisoning programs."[10] The ranchers and their Congressional allies did not demand or even mention "extermination." In private this remained their goal. In public they spoke of reasonable control and stressed that a properly conducted predator control program would save money and jobs without sacrificing wildlife.

Despite the Leopold report and the hearing there was little prospect of change. Udall's appointments and the new policy statement had given the program breathing room; critics would have to wait and see if the new policy worked. Even if there were no change, how could opponents force action? Western senators and representatives would keep Congress from doing anything, barring some enormous public outcry. The Secretary of the Interior was unlikely to take the lead. He relied on wildlife organizations, but also on western support in Congress and in the field. The president could, of course, ban poisoning, at least on federal lands, but why should he? The political dangers were large and apparent, the benefits invisible.

On the other hand, the program's defenses were a shambles and its opponents were stronger, more organized, and more militant than ever before. The ranchers had relied on public acceptance or indifference. That was gone. The PARC had justified its work as scientific. Now scientists were attacking the program with the accumulated research of a generation of wildlife studies. Worse, the militant humane societies were not amenable to the tactics the Service has used on the older wildlife organizations. With them it had appealed to common interests and common programs. The new humane groups were antihunting and anti-fishing and they rejected the common goals of wildlife management. The PARC might last, but it would only be by adopting new tactics and strategies.

The formation in 1954 of two new humane organizations may be taken as the handwriting on the wall. The Society for Animal Protective Legislation was something other than the "normal" humane organization; it gave the movement a continuing presence (however small) in Washington, lobbying on the federal level. The splitting off of the Humane Society of the United States (HSUS) showed another change in the movement. The HSUS left because a faction felt that the older American Humane Association (AHA) (founded in 1877) was not sufficiently "radical." The HSUS was against hunting and willing to say so. It was not a new stand, but it showed that the humane people thought the public was "ready" for a radical argument. A few years later another group appeared that was solely for wildlife. Some of its old adherents revived the Anti-Steel Trap League under the name Defenders of Wildlife. It quickly grew into a national organization with national goals, a journal (*Defenders*), a Washington office, a permanent staff, and the backing of a number of wildlife scientists, including the National Park Service's Victor H. Cahalane.[11]

The movement was developing its philosophical and ideological base. Pain was the central concern, and by the 1970s humane advocates would have pushed Bentham's pleasure-pain calculus to its logical conclusion—all animals, all pain. "If a being suffers," Peter Singer wrote in *Animal Liberation* (1975) "there can be no moral justification for refusing to take that suffering into consideration. . . . [T]he limit of sentience . . . is the only defensible boundary of concern for the interests of others."[12] Others would, by the 1980s, go beyond even that, attempting to construct ethical systems that took into account all forms of life and even landscapes.[13]

Humane advocates drew parallels between opposition to their cause and other forms of "prejudice and discrimination based on an arbitrary characteristic like race or sex." They saw "speciesism" as analogous to racism or sexism. On what grounds, they asked, could we distinguish between animals and humans of limited capacity or with damaged brains except on the "bare-faced—and morally indefensible—preference for members of our own species?"[14] Animal rights has come to justify actions as diverse as protests against raising chickens in confinement, freeing experimental animals, and destroying laboratories. In the 1950s things had not gone quite so far. Protests were limited to petitions and letters, action to demands for legislation. Still, they had gone far enough. Poisoning half a horse carcass to kill wild animals was unacceptable—and many people not allied with the movement agreed.

The humane movement depended, as it had before, on women. More than half the members were female, and women had, as they did not in the rest of the wildlife movement, significant representation on boards of directors and in other responsible positions.[15] The movement also reflected the demographic and social changes that had occurred since World War II. The young, the city dwellers, and the college-educated were disproportionately likely to agree with humane goals and join humane societies. A generation raised in the city and educated in an urban culture was breaking its ties with a rural society that still accepted, even celebrated, hunting, fishing, and the "conquest" of the wilderness.[16]

In the 1950s the wildlife wing of the humane movement developed its own agenda. One of its first causes was the protection of the West's wild horses. People were turning them into dog food, and a few people set out to make opposition to this activity a national cause. Wildlife managers wanted to remove introduced species, such as the burros in the Grand Canyon. Wildlife advocates took up that cause. Poisoning was another enemy, and one they concentrated on. In the late 1960s these groups also revived the anti-trapping campaign, and by the end of the decade it would be a major battleground. Since 1968 there have been almost 350 anti-trap bills introduced at various levels of government. There have been attempts in thirty-three states to ban the leghold or other "inhumane" traps. Efforts to ban hunting, previously the dreams of a few radicals in the humane movement, became extensive and respectable, and sportsmen's organizations and hunting magazines began to take them seriously.[17]

The Fish and Wildlife Service could neither ignore the new movement nor easily work with it. With the Audubon Society and the National Wildlife Federation, to take two of the largest "conventional" wildlife groups, it had common goals and shared values. Audubon had been cooperating with the U.S. Fish and Wildlife Service and the Canadian Wildlife Service since 1948 on a program to save the whooping crane. NWF's interest in duck regulations went back farther than that. Defenders of Wildlife, to take an example from the humane groups, had no ties to Fish and Wildlife and saw little advantage in forming them. The older organizations had almost as much trouble. The new groups put pressure on the Service, and they served as a foil for other organizations. With Defenders constituting, so to speak, the "lunatic fringe," the National Wildlife Federation seemed the voice of moderation, whatever it demanded. On the other hand, these organizations would not compromise or settle for changes that environmental groups regarded as effective solutions. NWF, for example, might settle for killing some predators if the program was selective, humane, and did not harm nontarget wildlife populations. Defenders insisted on ending all poisoning, and some of its members wanted to end all killing of wildlife.[18] The battle to ban poisons, then, was less a matter of "them and us" than of shifting interests and alliances. Each player used the others to achieve goals no one else fully shared.

The law was changing, and that too affected the Fish and Wildlife Service's ability to make and carry out policy. Judges were recognizing that bureaucratic and administrative decisions were often made without public input, and they began opening to the public agency decisions and the process by which they were made. They interpreted standing (the right to sue) more broadly. They expressed greater willingness to review agency decisions. They undertook to supervise in detail the implementation of their decisions. This was not particularly related to the environmental movement, but it affected wildlife policy. Agencies found that protestors could get a hearing in court and were willing to go there. Congress followed the same route, but it did what the courts could not: create new agencies to take on new areas—including environmental policy.

The passage of the National Environmental Policy Act (NEPA) and the formation of the Environmental Protection Agency (EPA) and the President's Council on Environmental Quality (CEQ) had a dramatic effect on the policy system that oversaw (and kept in place) the program of predator poisoning. NEPA established environmental protection as a national goal and set standards and procedures to reach that goal. The most important, for our purposes, was Section 102. It required that any federal agency prepare an environmental-impact statement before undertaking any major project. The statement would show how the proposed actions would affect the ecosystem. The section did not define any of its terms, and it produced an astounding amount of litigation over who had to produce a statement under what conditions and to what specifications. Even without those decisions environmental groups found Section 102 very useful. It gave them early warning of important projects and, by suits or the threat of suits, a way to affect agency actions at an early stage. The EPA and the CEQ were even more useful. They provided a way around the woolgrowers' Congressional roadblock.

By the time CEQ and EPA were organized in the summer of 1970, patience with the new Division of Wildlife Services was wearing thin. Seven years after the Leopold report and four after the PARC's adoption of "Man and Wildlife" as its policy, nothing seemed to have changed. Protests were even beginning to appear outside the restricted world of *Defenders, Audubon*, and *National Wildlife*. In June 1970, *The New Yorker* published Faith McNulty's article on the black-footed ferret (which became the book *Must They Die?*).[19] The following March *Sports Illustrated* (not a major environmental journal) published a three-part series on the "Poisoning of the West" taken from *Slaughter the Animals, Poison the Earth*.[20] They gave a vivid, if not balanced, picture of the program. There were descriptions of deaths and near-deaths from people stumbling on "coyote-getters," and reports of stockmen setting coyotes on fire or sawing off the animals' lower jaw and turning them loose or turning dogs on them.

The same month these articles appeared Defenders of Wildlife sued the Fish and Wildlife Service. It claimed that Section 102 of NEPA required an environmental-impact statement for the predator and rodent-control program. In May 1970 poisoning became a national issue. Boy Scouts found several dead eagles in a canyon near Casper, Wyoming. A search by the local chapter of Audubon turned up two dozen carcasses, both bald and golden eagles. The birds had been poisoned by feeding on sheep carcasses heavily laced with thallium salts. In June a Senate committee held hearings on the incident.

At the hearings Fish and Wildlife Service officers told the subcommittee that the agency had ended all use of thallium compounds some years before. They were too inhumane. The Service had to keep its distance from the incident, for testimony suggested there was a war on wildlife in Wyoming. Stockmen around Casper had purchased sixty-five pounds of thallium salts in the previous two years. A Colorado firm, the Craig Wool Warehouse Company, had purchased 120 pounds of thallium salts (the only legal use was for alloying and metal working). Sheep carcasses found in the region had enormous amounts of poison. One had had four pounds of thallium, where four ounces would have produced a lethal dose in the meat. Thirty-two dead sheep were found with eight times the concentration of poison needed to kill coyotes, another twelve had sixteen times too much.[21] An airplane pilot testified (under a grant of immunity) that he had worked for several months flying in pursuit of coyotes and eagles, which his passengers had then shot. He knew the toll was more than five hundred bald and golden eagles. He also told of seeing Wyoming game officers pass, without comment, a pile of over sixty eagle carcasses.[22]

A second set of hearings in December generated more publicity. Both sides took the chance to make their cases again. The stockmen had affidavits from people who had seen eagles preying on sheep. They produced statistics showing high, and rising, losses to predators. The National Woolgrowers' Association and its state branches came to their members' defense. Congressmen pleaded for their constituents. Predation, they argued, was in many areas so serious that even good ranchers needed federal help to stay in business. Their opponents challenged the woolgrowers' statistics, economic assumptions, and biological knowledge. Oliver Scott, an Audubon member from Casper, Wyoming, pointed to several studies of eagles and sheep which showed that the birds were not a significant problem. One 1968 study, sponsored by the Audubon Society, the National Woolgrowers' Association, and the Fish and Wildlife Service (then operating as the Bureau of Sport Fisheries and Wildlife) estimated that eagles took between six one-hundredths and three-tenths of one percent of the lamb crop.[23] Charles Callison of the National Audubon Society pointed to a letter from an extension specialist in New Mexico. It said that losses to predators were serious, then asked the rancher about his own. With such prejudiced studies, Callison asked, how could anyone accept the ranchers' figures?[24]

Cleveland Amory, president of the Fund for Animals and a member of Defenders of Wildlife, attacked the entire poisoning program, using as his text Jack Olsen's *Slaughter the Animals, Poison the Earth*.[25] His testimony was a marked contrast to earlier humane objections to the PARC program. Those had been against cruelty; his were against killing. The humane societies were coming into the open. The public, they thought, was ready for a radical defense of animals. The appearance of several congressmen to speak against poisoning suggests that they were right. Senator Gaylord Nelson of Wisconsin called for a national policy on predators and an end to predator poisoning. He had, he said, introduced a bill that would do just that. Senator Birch Bayh testified about his own anti-poisoning measure, introduced in June. Legislators had looked at the conduct of the program before—the authorization for the Biological Survey's ten-year plan in 1930 had brought some sharp questions—but never had they attacked the program or the standard means used to carry it out. Still less had they introduced bills to ban a basic part of the predator control "arsenal."

Into the Bureaucracy

Public indignation put pressure on the program but bureaucrats did the work. The steps leading from public outrage to a ban on poison are not always clear—agency files are not yet open, some people will not talk, and others, having talked, deny their statements. It seems certain, though, that the chief scientific advisor for CEQ, Lee M. Talbot, played a crucial role. Talbot, a biologist interested in African wildlife, had been concerned about the predator control program even before he went to work for CEQ, and he moved quickly to get a new review of the program. The eagle killings and Defenders' suit gave him leverage.[26]

He must have made effective use of it, for a month after Defenders went to court, CEQ and the Department of the Interior announced a new joint review of predator control policy. In July 1970, just after the first eagle hearings, Secretary of the Interior Rogers Morton announced the makeup of the new committee. It was composed entirely of biologists. Two of them, Stanley Cain (the chairman) and Starker Leopold, had been on the 1964 committee. The others, the woolgrowers immediately complained, were known to be hostile to predator control.[27] Their objections and Interior's response are a measure of how much ground the woolgrowers had lost. In 1964 there had been no obvious concern that a review of predator control would change the program. Now the woolgrowers feared they would be ignored. They wanted a representative of their interests added to the committee. Earlier the Secretary of the Interior probably would had acceded; all Morton promised was that the Cain committee's report would not be used as the basis for policy changes until everyone had a chance to respond to it.

The committee's charge was to review the program and assess the extent to which the recommendations of the Leopold committee had

137

been implemented. It concluded that too little had changed. This, it said, was not the fault of the people in charge. They had tried, but "[i]t has been impossible for the Wildlife Service to enforce its guidelines in the field." The reasons, which the committee spelled out in detail, boiled down to organization and funding. There was a "high degree of built-in resistance to change in the basic machinery" of the program. Current field agents were the same people who had carried out the old policy. They were friends and neighbors to the sheepmen and had been accustomed to measuring results by the numbers of coyotes killed. They continued to do so. Funding was another key element. That the cooperators paid for part of the work "maintains a continuity of purpose in promoting the private interest of livestock growers." Money served "as a gyroscope to keep the bureaucratic machinery pointed towards the familiar goal of general reduction of predator populations, with little attention to the effects of this on the native wildlife fauna."[28]

The committee wanted the program completely overhauled, though it acknowledged that such a course would require "substantial, even drastic, changes in control personnel and control methods, supported by new legislation, administrative changes, and methods of financing."[29] Congressional and state appropriations, it said, should replace the cooperator's funds, and wildlife biologists should replace the untrained trappers. The goal of a general reduction of predators should yield to a plan that emphasized removing only the offending individual animal. There should be an end to aerial hunting, aerial broadcasting of poison baits, and rodent control in areas where secondary poisoning could occur. Predator control should move away from poison, emphasize a long-range plan of scientific research, and be more sensitive to the problems of endangered species.[30] Short of putting a tax on sheep to support old coyotes, it is difficult to think of a set of recommendations that would have angered the ranchers more.

Critics, then and later, have complained about the composition of the group, its speedy judgment, and the scientific evidence it used in making its recommendations.[31] Since the report was the basis for the ban President Nixon declared on predator poisons three months later, it is worthwhile looking at these issues. The committee's charge was to assess the extent to which the Division of Wildlife Services had implemented the reforms recommended by the Leopold committee and adopted in the policy statement "Man and Wildlife." Wildlife biologists were the only appropriate people for this task. However, CEQ and Interior could have found wildlife biologists who supported the existing control program. That they did not suggested they were not looking for them. Some people thought this was the case—that Assistant Secretary Nathaniel Reed, Secretary Rogers Morton, and Talbot wanted to halt the use of poisons and chose a committee that would recommend it.[32] The speed with which the group worked has been taken as an indication that it had an end in view. But the members were all familiar with the situation; they did not need much time to reach a decision.

The nature of the case has also been used to support charges that the result was predetermined. The Cain committee, it is said, had little more scientific evidence than its predecessor in 1964, and it was not an open-and-shut case against predator control or even poisoning. The committee had gone beyond science to its own opinions and values. This is probably true. Despite a generation of wildlife research, biologists are rarely able to provide unambiguous answers to complicated questions about ecosystems and populations. The charge, though, is a bit disingenuous. Scientific evidence had not been responsible for starting or continuing the program. Western congressmen had pressed for it as a service for their constituents. Once in place, the interests of the woolgrowers and the agency had kept it going. Now a large part of the population had come to feel that "varmints" were valuable and should be preserved.

The woolgrowers' charges against the process by which the report was made into policy are well founded. They were ignored, shut out, and never given a chance to comment. The whole thing, from the time the committee turned in a draft to Nixon's speech banning poisons on federal land, was unfair and undemocratic. It violated all the standards by which policy is supposed to be made in a democracy. To this the anti-poisoning forces would doubtless reply with mutterings about the goose and the gander. Perhaps the best judgment is one a New Zealand biologist made about the bureaucratic politics of his own country's deer policies. "Real politics," Graeme Caughley said, "are for consenting adults in private. We are a little embarrassed by them and wish they did not parallel so precisely our own grubby, but less publicized, personal goings on."[33]

Even in this imperfect world the Cain report might have gone through the entire process of hearings, debate, and discussion before anything was done with it. That it so quickly became the basis for a new policy was the result of the situation in 1972. It began with Talbot's handling of the report. When he got it in November 1971, he circulated it under tight security in the Department of the Interior and the Office of Management and Budget. This "dry run" allowed him to assess opposition within the administration and tailor recommendations to meet it—without giving the woolgrowers an opportunity to comment. They did not, in fact, even realize the report had been written.[34] Recommendations based on the report then went directly to the White House.

Ordinarily, aides would have sounded out potential opponents and interested parties. They would have taken political sounding. The Nixon White House lacked this give and take (on which Talbot may have counted). Secretive and suspicious, Nixon had arranged his lines of communication to minimize input and restrict access.[35] Politics also played a part. The campaign of 1972 was just around the corner and the environmental vote promised to be important (it was less than two years after Earth Day). Nixon had backed the supersonic transport plane and the Alaska pipeline. A ban on predator poisons, announced with suitable fanfare as part of a major speech on the environment, might defuse the opposition. It would, in any event, be a safe decision. The woolgrowers would protest, but they were unlikely to vote for a Democrat in the fall, particularly for a liberal.[36]

On 18 February 1972 President Nixon did make an environmental State-of-the-Union address, setting out goals for his administration. One was an end to the common predator poisons—Compound 1080, strychnine, and cyanide. Nixon announced an immediate end on the use of these compounds on federal property and said he would seek legislation to extend the ban to private lands. He promised that the policy of general reduction of predator populations would yield to a new one of removing only the offending individual animal. A month later the Environmental Protection Agency banned the interstate shipment of cyanide, strychnine, 1080, or thallium for predator control. In October, Congress passed the 1972 Federal Environmental Pesticide Control Act, requiring the states to conform to federal standards.[37]

The woolgrowers were outraged—with good reason. Whatever its motives and whatever the evidence, the administration had made a major decision affecting their lives and livelihoods without even the pretense of seeking their views or taking their needs into account. The National Woolgrowers' Association and its state branches protested to Congress, the president, and the Department of the Interior. It was an uphill battle. Nixon may have made his decision for immediate political ends, but many people agreed with it. Ten-eighty was only accidentally a casualty of Nixon's reelection strategy. There was a widespread desire to replace destructive methods of dealing with the natural world with ones that preserved and protected it. The woolgrowers were victims of more than environmental politics. They stood on the wrong end of a shift in American culture.

The shift was more complicated than it appeared from this fight. Ten-eighty had been a convenient target for humane groups and environmentalists because it presented what seemed to be simple choices. Now more complicated questions had to be faced. This was most apparent in the fight, then going on, over the enactment and enforcement of a comprehensive program of protection for endangered species—one of which was the wolf. Here, even more than in coyote control, the rules of the game and the distribution of power within the system were changing rapidly, and the fight would test the strength of Americans' commitment to environmental values.

CHAPTER TEN

Saving Species

*The beauty and genius of a work of art may be reconceived, though its first material
expression be destroyed; a vanished harmony may yet again inspire the composer; but
when the last individual of a race of living beings breathes no more, another heaven and
another earth must pass before such a one can be again.*—William Beebe, *The Bird,
Its Form and Function*

Since the extinction of the dodo, humans have been aware that they
could exterminate a species, but for most people the danger was as
exotic as the dodo. The extinction of the passenger pigeon and the
near-death of the buffalo unsettled many people, but it brought no
movement to save species. There were efforts to save some of the most
beautiful, useful, or historic animals—egrets, herons, buffalo, or antelope—but public attention rarely strayed beyond the larger birds
and mammals. Not until after World War II, when economic development and habitat destruction made it a real and present danger, did
extinction become an important issue. Then the public's new appreciation of ecosystems fueled the movement and gave it a new direction.
Now people wanted to save all species and to save them as part of a
system. All, they argued, have a place and a purpose on earth. In a
decade-long burst of enthusiasm, Americans wrote their ideals into
law. Congress passed endangered species acts in 1966, 1969, and
1973. It passed an act for whales, seals, and their relatives: the Marine
Mammal Protection Act of 1972. The Convention on International
Trade in Endangered Species of Wild Fauna and Flora provided the
basis for a system of international protection for rare species.[1]

The program thus established was different not only in scope but in
kind from earlier efforts. In the past, people had focused on a few
kinds of wildlife, which they had protected by setting aside refuges
and closing hunting seasons. Now virtually all forms were to be saved.
There was an elaborate set of regulations and a continuing program
of scientific research and restoration. For wolves the change was dramatic. Bounties had been common; but classification of wolves under
the Endangered Species Act of 1973 did away with them. Wolves, out-

side Alaska, had been rare; the 1973 law proposed to re-establish the species in as much of its former range as possible. Americans' new enthusiasm for a pristine wilderness, with all its species, had helped change the wolf's image, and now environmental enthusiasm was changing the animal's legal status as well.

Laying Foundations

Like the struggle over predator poisoning, protection for endangered species involved politics, mainly bureaucratic politics. Not surprisingly, the Fish and Wildlife Service (its official title from 1957 to 1974 was the Bureau of Sports Fisheries and Wildlife, but let's keep things simple) bore the brunt of this enthusiasm. The agency had been working in the area of preservation since the 1930s, when it had joined the National Park Service in efforts to save the trumpeter swan. There had never been, however, the kind of systematic program these laws envisioned. The agency had only begun to make a start on this in 1962, when it formed a Committee on Endangered Species in the Division of Wildlife Research. The committee had two major tasks: listing all endangered species, and deciding what the agency could do to save them.

The second was the easier. Some species could be helped by setting aside refuges (the simplest case). Others were scattered over a larger area and would require more resources (an important consideration for a bureau with limited resources). Finally, with some cases it was not clear if the species was in danger of extinction.[2] In a few cases, species were already being helped. The trumpeter swan was one, the Hawaiian goose another. A third was the whooping crane, which the Fish and Wildlife Service, the Canadian Wildlife Service, and the Audubon Society had agreed to work on cooperatively in 1948.[3] Most species, though, had no formal protection, and there were no organized efforts to aid them.

Listing of endangered species was a more difficult matter. Though there were some obvious cases—like the whooper—there were little basic biological data for many species. Accurate censuses, particularly for animals that did not congregate in a small area at some period of the year, were scarce. In 1964 the committee, now the Office of Endangered Species, had compiled a list of some sixty species that it was certain were endangered. They included birds, mammals, fish, reptiles, and amphibians. Among them were the black-footed ferret, the wolf, and the red wolf.

The prospect of another part of the Service scrutinizing its operations triggered the Division of Wildlife Service's bureaucratic reflexes. It could not come out against saving species, but it could object to parts of the program. And it did. It protested, for instance, that the timber wolf, listed as *Canis lupus lycaon*, was doing well in Canada. It was not really endangered. The name, in any case, ought to be changed to Eastern timber wolf to avoid confusion (a pencil note in the margin of the memo says "no change"). The black-tailed prairie dog, it said, was "plentiful" (the margin here says "rare").[4]

While this low-level bureaucratic skirmishing was going on, public interest was forcing Congressional action. Environmentalists, wilderness advocates, and wildlife protection groups (now including humane societies) saw in extinction an obvious, immediate danger.[5] All turned to the federal government. Though each faction had its own agenda, they shared certain assumptions. Everyone thought that all forms of life—mammals, birds, fish, reptiles, insects, mollusks—should be preserved. They wanted them saved not in isolation but as part of a connected system. Zoo specimens were not enough. There had to be wild populations with all the behavioral characteristics that had evolved as part of the creatures. Finally, all agreed that endangered species protection should be backed by extensive federal research on the ecology of each species that would guide plans to restore these creatures to their status of active parts of the ecosystem.

Things began modestly, with a provision in the 1964 Land and Water Conservation Act that authorized the Secretary of the Interior to purchase land for the preservation of native wildlife and game. It was an afterthought, not the beginning of a program. Congress did not even define "native wildlife," much less outline preservation programs. When Congressional appropriations committees complained that the provision gave the Secretary new powers more appropriately conferred by statute, the administration submitted the bill that became the Endangered Species Preservation Act of 1966. It directed the Secretary of the Interior to use his existing authority to purchase and manage land for a new purpose—protecting native wildlife threatened with extinction. It appropriated $15 million from the Land and Water Conservation Fund for land purchases and enjoined the Secretaries of Interior, Agriculture, and Defense to protect endangered wildlife on lands under their jurisdiction. There were limits. Federal agencies need consider wildlife protection only when it was consistent with the primary mission of their agencies. Only $5 million could be spent for land in any given year and no more than $750,000 on a single project.[6]

This was more symbol than substance. There was no new federal power to protect wildlife, and there were no restrictions on taking or commercial use. The Secretary of the Interior had no authority to take action on private land. The act did not define endangered species and only sketched in the process by which they were to be declared in danger. Formal procedures were vague. The Secretary was required to protect native wildlife threatened with extinction. He was to make his decisions in consultation with the affected states and with the advice of "interested persons." He was to publish his findings in the *Federal Register*. That was all. States had most of the wardens and resources, and their cooperation was essential, but on this subject the act said only that the Secretary should cooperate with the states "to the maximum extent practicable."

Wildlife's defenders wanted more: standards for listing species, penalties for violations, and, most important, worldwide protection, enforced by controls on commercial traffic. A law that did this would give the Service new administrative responsibilities. Like all other bureaucrats, Service officers shuddered at the thought of letting others decide how and when and where they should do their job. To their own enthusiasm for wildlife they now added the desire to get an act that would work and that they could work with. In 1967 Representative Dingell introduced a new endangered species bill—written by the Fish and Wildlife Service. In August 1968, it passed the House without opposition. It met resistance in the Senate. Furriers were afraid that the law would harm their business. They learned late of the House's action, but they learned. Industry representatives rushed to Congress. They had no cases to cite—they couldn't have under the circumstances—but they raised enough objections to stall Senate action. The session ended with no bill.[7]

Dingell and three cosponsors reintroduced the bill early in 1969, and wildlife advocates and industries using wild animals spent the year arguing over the nature and extent of the protection the United States should offer wildlife abroad. Furriers objected to controls on imports and exports, but even more to American determination of what species were to be classified as endangered. Unilateral action, they argued, would be counterproductive. It would not protect animals; it would only move jobs and profits out of the country. They called for an international agreement. Wildlife, humane, and environmental groups insisted on listing by the United States. They did not oppose an international agreement but thought America should move first and quickly. Others, they said, would then follow.

Because the bill affected people's lives and jobs, it produced far more controversy than its predecessors. In 1965 the Wildlife Management Institute had been the only group to come to House hearings, and only four others—the Audubon Society, Defenders of Wildlife, the National Wildlife Federation, and the Sport Fishing Institute—had submitted statements for the record. In 1969 thirty-nine groups sent witnesses at the final House hearings. They ranged from the Animal Welfare Institute and the Sierra Club to Gators of Miami and Brooklyn Better Bleach, Inc.[8] They were a different mixture as well. Business and humane societies took a much larger part in the proceedings.

In 1969 environmental enthusiasm was high enough to overwhelm the objections of the furriers, trophy hunters, and other opponents, particularly since they could do little more than point to possible problems that might arise if the bill passed. This was not very persuasive, and Congress passed the Endangered Species Conservation Act of 1969 almost unanimously. The bill put in place a stronger program. The Secretary of the Interior received authority to list species threatened with extinction and to control their passage into and out of the United States. "Wildlife" now included crustacea and mollusks. The new law amended the Lacey and Black Bass acts, increasing their range. The process for listing species was placed under the provisions of the Administrative Procedures Act. There was also more money: up to $2.5 million from the Land and Water Conservation Fund could now be spent on a given project.[9]

The new law, however, applied only to foreign wildlife, and, since the controls were on shipping, only to those species that were sold. The Secretary could not restrict trade unless a species was threatened with worldwide extinction. He could not, in short, act until the situation was desperate. Wildlife people thought the act useful but not sufficient. They wanted legislation that would protect native wildlife, keep species viable, aid in the recovery of those threatened with extinction, and protect ecosystems. They wanted more.

A Changing Situation

These laws did not provide for federal wolf management, but public concern was beginning to affect plans made by the states. The species, paradoxically, was in less trouble than it had been a generation before, when few had been concerned about the wolf. The greatest pressure for extermination had come from livestock interests, and as agriculture and grazing in North America reached the limits set by climate and soil, pressure for "control" declined. The wolf population in North America reached its nadir around 1940. Since then it had been stable and in some areas it had even increased.[10] The now popular conviction that wolves were precious parts of our vanishing wilderness and the desire to restore the species made people more sensitive to the issue.

The only large numbers of wolves in the United States were in Alaska, and it was here that public concern first affected wolf management.[11] In the 1950s the territory had cooperated with the PARC, but by the end of the decade it was abandoning the idea of reducing wolf populations to save game. It began limiting control to those situations where wolves had been shown to be causing problems. It closed some areas to wolf killing entirely to allow research on predator-prey interaction. Statehood brought local control of wildlife, but the legislature did not reverse this course. It declared the wolf a furbearer and a big-game animal (making wolf killing subject to trapping and hunting laws). In 1968 it allowed the Fish and Game Department to end bounties on wolves in individual Game Management Units. The department quickly did so, leaving them in force in only a few areas in the southeastern part of the state.

Changing public ideas had an even greater impact on management of the several hundred wolves that lived in the Superior National Forest and in the Boundary Waters Canoe Area in northeastern Minnesota. They were the remnants in the United States of the Eastern timber wolf (*Canis lupus lycaon*) whose packs had once roamed from Hudson's Bay to the Carolinas and from the Atlantic coast to the edges of the Great Plains. In the 1930s there were still a few stragglers on the Upper Peninsula of Michigan and in northern Wisconsin. By the time Americans began to be concerned about wolves, the only breeding populations—the last wolves in the continental United States—

147

were those along the Canadian border and scattered through the northern third of Minnesota.[12]

People in the state were deeply divided. The wolves' range was a checkerboard of private and public lands, and farmers, particularly those with stock in pastures carved out of the forest, strongly opposed wolf protection. So did many hunters, who were convinced that wolves killed "their" deer. The state also had a long tradition of conservation and wilderness appreciation and a strong pro-wolf movement. Older wildlife organizations combined with new ones, some formed expressly to save the species—like HOWL (Help Our Wolves Live)—to push for full protection. Calls for saving the wolves and for an appreciation of the state's wilderness heritage clashed with demands that the state protect its citizens and their property. Tempers rose along with interest; one letter to the editor condemned the "do-gooders and the right-to-live radicals" who had stopped "on the verge of success" the efforts of "Minnesota taxpayers, State Conservation officials and residents of northern Minnesota" to exterminate the timber wolf.[13]

Wolves became a political issue. In 1966, after a bitter fight, the legislature voted to continue paying a bounty on wolves. Governor Karl F. Rolvaag vetoed the bill. This earned him the warm congratulations of Secretary of the Interior Stewart Udall and the enmity of many Minnesota outdoorsmen. Two years later the district office of the Fish and Wildlife Service reported that wolf policy might play a part in Minnesota's forthcoming election. One candidate, it said, was preparing a resolution endorsing protection of all endangered species in Minnesota.[14]

The public was becoming more sympathetic to wolf protection. One indication was reaction to a television special aired in November, 1969, "The Wolf Men." It was the story of modern methods of wolf killing, complete with pictures of trapped wolves, gunners killing wolves from airplanes, and a hunter eating raw wolf meat from one of his kills (he said later he had done it as a joke for the cameramen). The program was controversial even before it was shown; anti-wolf organizations in Fairbanks got an injunction blocking its broadcast in Alaska. The rest of the country saw it and "besieged [the Fish and Wildlife Service] with telegrams, telephone calls, and letters." Reaction extended across the border. A Fish and Wildlife staffer said that a friend in Ontario "told me that . . . citizens almost snowed his office with letters commenting on the program." It was "[d]emocracy in action. . . . I'm sure we all recognize that they are our clientele."[15] The clientele, though, was divided, and the Service had to work out plans acceptable to everyone.

Its primary need was for information. To restore wolves or even to manage existing stocks, the agency had to know how many wolves it took to make a pack, how far they ranged, how much land they needed, and how fast the population grew. Now the studies begun in the 1950s began to pay off. In 1957 the National Park Service had entered into an agreement with Purdue's Durward Allen for a long-term study of wolves and their prey on Isle Royale. One of the first students there was L. David Mech. His study, *The Wolves of Isle Royale*, appeared in 1966 as a monograph in the same series—*Fauna of the National Parks of the United States*—that had published Adolph Murie's *The Wolves of Mt. McKinley* some twenty years earlier.[16] Allen continued work on Isle Royale, and on the mainland other scientists joined in. In 1965, a year before Mech's work appeared, Douglas Pimlott wrote to the head of the Canadian Wildlife Service that he had decided to concentrate his work and that of his students on canids. He outlined an ambitious research program, one part of which was an intensive study of the ecology and population dynamics of wolves, and the Ontario wolves were just across the border from Minnesota.[17]

The studies were in large part the result of the new technology of radio tracking. Scientists had begun experimenting with this technique in the early 1960s, and it quickly became apparent that radio collars could do as much for large-mammal studies as rodent traps had done for small-mammal work in the 1870s or bird banding for migration in the 1920s.[18] It was particularly valuable for species like the wolf, which ranged over large areas at a quick pace. Now scientists could follow the animals, plot their home range, and determine patterns of use and times of activity. They could study dispersal and colonization and find out how stable packs and territories were. In October 1968, the Fish and Wildlife Service made a grant of $3,200 to L. D. Frenzel and L. David Mech for the radio tracking of two adult wolves. It was the beginning of studies that would lay the groundwork for management and restoration of wolf populations in the United States.[19]

The Service also used older techniques to answer other questions. Chief among these was the taxonomic classification of the canids of North America. This was not a theoretical concern. The Division of Wildlife Services had already used taxonomic arguments to protest the initial listing of the eastern wolves. The state of Minnesota had made the same argument when its wolves fell under the 1966 Endangered Species Act. There was also the red wolf. If it was a subspecies of the timber wolf, its treatment and status would be different than if it was a distinct species. Any program to save it first had to classify it—and defend that decision.[20] There was also the question, less urgent, of the eastern coyote. What was its relation to wolves and to domestic dogs? A related problem was the distribution of wild canids in North America, their populations, the condition of their ranges, and the nature and extent of human disturbance of these areas. The Service had to know the situation on the ground.

In 1973 the Fish and Wildlife Service and the New York Zoological Society hired Ronald Nowak (a student of E. Raymond Hall) to review the status of the wolf in North America. Nowak, who undertook a revision of North American *Canis* for his Ph.D. dissertation, was in a position to answer many of the Service's questions about canids (he is now a staff specialist with the Office of Endangered Species, working on mammals). The answers he provided became part of the information base used in the late 1970s to guide policy.[21] Classifying groups by sorting out the accumulated information and interpreting or reinterpreting it in the light of new evidence and ideas is a continuing process. Taxonomy, though it appears to us in high school and college biology as a dead (and dull) operation, is in fact still quite alive.

A generation earlier Stanley Paul Young and E. A. Goldman had classified and described the wolves of North America.[22] Nowak had more material, a new problem (the eastern coyote), and, in multivariate analysis, a new tool to deal with hard cases and incomplete material. He could much more easily compare specimens with regard to many different characteristics, separating what were, by older practices, blurred and overlapping populations. His general conclusions, not surprisingly, were close to those of earlier workers. He recognized three species of living canids in North America: the wolf, *Canis lupus*; the coyote, *C. latrans*; and the red wolf, *C. rufus*. Several subspecies of wolf were now extinct, he said, but the remainder seemed to have stabilized. The eastern coyote he recognized as a variant of *C. latrans*. Specimens taken from the 1930s and later showed some interbreeding with wolves. This, Nowak thought, might account for the survival of the animals in conditions so different from the coyote's usual haunts.

The animals, though, were still coyotes, clearly distinguishable from wolves.

Distribution was somewhat different than it had been a generation earlier, particularly for the coyote. Despite poisoning, the animal was flourishing in the West and expanding its range in the East, working its way down the chain of the Appalachians. The red wolf was in trouble. Trapping, hunting, and human disturbance of its remaining habitat threatened it with extinction. Only a remnant population remained in the Gulf coast swamps straddling the Louisiana-Texas border, and in these areas human disturbances had begun to break down the behavioral mechanisms that isolated the species. A "hybrid swarm" of mixed coyote-red wolf ancestry threatened to breed the "pure" red wolf out of existence.[23]

Full Protection for Endangered Species

Under the circumstances, even experts might differ on the best course of action or the status of a particular population. Americans, though, did not want discussion of complex questions. They wanted to save endangered species. Their first goal was to strengthen the 1969 Endangered Species Conservation Act. They wanted more protection for domestic species and foreign ones alike, and a more restrictive "economic hardship" clause. Wildlife advocates charged that the provision, designed to provide relief for those the act had caught with large stocks of products now barred from commerce, was being abused. The government was routinely granting applications for exemptions. The law also called for an international conference on wildlife protection, to be held by 30 June 1971. There was no visible progress toward that end.

Conditions favored a new wildlife law. Environmentalism had become a popular crusade. Leopold's *Sand County Almanac* was now a paperback in Ballantine Books' "Style of Life" series, recommended in the *Whole Earth Catalog*. Bumper stickers proclaimed "Ecology Now!" The environmental sins of corporate "Amerika" were a standard part of the counterculture's indictment of capitalism and imperialism. It was not just the counterculture, though. Many middle-class, "average" Americans were concerned. In April 1970, several hundred thousand people rallied in Washington for a celebration of "Earth Day." Schools around the country held "teach-ins" and special programs on the environment. Politicians who had previously shown little concern with natural resources, wildlife, or outdoor recreation hastened to find a place under the new banner. President Nixon's declaration that the 1970s were the decade when America must restore "the purity of its air, its waters, and our living environment" appeared on the back of a paperback edition of *Silent Spring*. Conservatives and liberals were uniting in a coalition to clean up the nation's air and water—and save its wildlife and wilderness.

In his environmental State-of-the-Union address Nixon called for greater protection for endangered species.[24] Less than a month later Representative Dingell and Senator Hatfield introduced bills (written by the Fish and Wildlife Service) to provide that protection. Neither passed, but a related measure that foreshadowed the Endangered Species Act of 1973 did. It was the Marine Mammal Protection Act (MMPA) of 1972.[25]

The MMPA was a drastic change in federal regulation of marine wildlife and ecosystems. Earlier statutes had been concerned with managing a single species for long-term economic gain, the "maximum sustainable harvest." This one covered all marine mammals; it had as its goal the preservation of the marine ecosystem; and it spoke of "optimum sustainable populations" and "the adverse effect of man's actions." Species "should not be permitted to diminish beyond the point at which they cease to be a significant functioning element in the ecosystem of which they are a part."[26] To achieve these sweeping ends MMPA preempted state authority over all species of marine mammals. It put an indefinite moratorium on their taking (very broadly defined) and imposed an elaborate set of rules on commercial traffic in these species and their products. It called for an ambitious program of international protection and made its policies the official American negotiating position in future international conferences.[27]

In the spring of 1973 the United States moved closer to a system of worldwide protection for all endangered species with the signing in Washington of the Convention on International Trade in Endangered Species of Wild Fauna and Flora (CITES). The agreement provided administrative and scientific machinery for saving endangered species. It has separate classifications for animals in immediate danger of extinction (defined in Appendix I and usually called "Appendix I species") and those (Appendix II) less threatened or so similar to endangered species that trade had to be restricted in order to protect rare ones. Each shipment of each species had to have import and export permits. All signatories agreed to appoint a national scientific authority to pass on population levels and numbers to be allowed in trade and a management authority to oversee the regulations. The convention provided as well for regular meetings to discuss listings, amendments, and enforcement.[28]

CITES was signed a week before hearings opened on the bill that become the 1973 Endangered Species Act and the first witnesses at the hearings were the American delegates to the treaty conference.[29] There was overwhelming public support for endangered species protection and the law passed by votes of 390-12 in the House and 92-0 in the Senate. That final harmony masked a bitter battle over the law's provisions. Affected industries, realizing that something would pass, sought in committee to minimize what was, from their point of view, the damage. Environmental and humane organizations wanted as comprehensive and rigorous a bill as possible. Enforcement was probably the hardest case, with wildlife groups calling for authority to go to the Secretary of the Interior, and businesses asking that the Secretary of Commerce enforce the law and the countries of origin decide which of its species were endangered. Then there were the definitions of "wildlife" and "taking." These were critical. If, for instance, "taking" was defined as an action that harassed or bothered an endangered species, it would mean considerably more than if taking meant killing or wounding the animals. A broad definition of "wildlife" would have similar effects. There was also strong pressure to include plants, and there were disputes over the wording of exemptions for economic hardship.[30]

Proponents of a strong act won on almost every issue. Following the MMPA, the Endangered Species Act of 1973 defined "taking" broadly enough to include almost any action that threatened or disturbed an individual of an endangered species. It went further: areas vital to a species' survival might be set aside as "critical habitat." On land so designated, almost any action would come under federal scrutiny. The law covered almost all kinds of animals—down to invertebrates, crustacea, and insects—and plants. Almost the only things left out were microscopic life and imported insect crop pests. Like CITES, the Endangered Species Act established separate lists for "threatened" and "endangered" species, and it allowed the listing of species similar to endangered ones. The Secretary of the Interior was responsible for most wildlife under the law. Private citizens, too, had a role. They could petition the Secretary to list a species and were entitled to a formal response. The act also eliminated restrictions on the use of the Land and Water Conservation Fund to acquire land for endangered species protection.

It had as its purpose the protection of "the ecosystems upon which endangered species and threatened species depend." Section seven directed federal agencies "to insure that actions authorized, funded, or carried out by them do not jeopardize the continued existence of such endangered species and threatened species or result in the destruction or modification of [critical] habitat."[31] Instead of measuring actions by immediate human needs and wishes it sought long-term land stability, even if that required some immediate sacrifices. It also provided a mechanism to restore species. The Secretary was to appoint a recovery team for each threatened or endangered species. Its job was to make plans to restore the species as a functioning part of the ecosystem. Far more than the acts that preceded it, the 1973 law wrote into the statutes the new values about nature and man's place in nature that had become popular since World War II.

The New Situation

The passage of endangered species acts and the banning of predator poisoning changed the relative positions of wildlife defenders and their opponents. Wildlife groups had embarked on a crusade, backed by public enthusiasm and alarm. They had conquered. Now they had to administer. Their opponents, once firmly entrenched, were now on the outside, sniping. Both groups had to change their tactics. For environmentalists and humane types, marches and press conferences were out; lunches with assistant undersecretaries, offices in Washington, and a close reading of the *Federal Register* were in. Their opponents had, in some ways, the easier task. They did not have to attack popular reforms but simply pointed to the consequences of particular actions. They could propose amendments, changes, and exceptions, and it was difficult for a movement founded on popular enthusiasm to counter this guerilla strategy. Modifications of an executive order, definitions, the disposition of small plots of land, and seemingly simple changes were important but not interesting.

The new laws and regulations were foundations, but no more than that. Enforcement and interpretation would determine how far Americans would pursue the idea of environmental restoration, what other values they would sacrifice for this one. It was not, of course, that simple. Policy was not a direct reflection of public values. Bureaucratic rivalries, existing agencies and mechanisms of enforcement, and the wishes of those in power also have an effect. What has emerged in the last decade is a consensus on nature protection and the place of predators on the land, but it is also a political compromise.

CHAPTER ELEVEN

Finding Equilibrium

1973-1985

I suppose never before in history did we reach a period when any civilized nation, or uncivilized, has prohibited its people from protecting their own animals with the most efficient methods that were available to protect them. Certainly never before has there been any widespread feeling that predators were the friends of mankind and were something to be propagated in preference to edible animals. But we are now faced with something of that situation.[1]

With this blast Representative W. R. Poague of Texas opened hearings on predator control before the House Committee on Agriculture in 1973. He exaggerated, but he expressed the woolgrowers' feelings. They had used government grass, and the government had helped them get rid of the "varmints" that killed their stock. They had been the conquerors of the wilderness, the bearers of civilization. Now they were careless destroyers of nature, and coyotes were precious things to be preserved and protected.[2] The wooolgrowers were fighting for more than a restoration of predator poisons; they wanted public approval.

That may have been beyond their reach, but in 1973 they could still hope to restore poisoning. Nixon had responded to the needs of the moment. When the moment had passed he might be persuaded to reverse himself. If he did not, there were other avenues of approach. Federal predator control officials would support the woolgrowers. Western representatives in Congress, sitting on oversight and appropriations committees, could press the Department of the Interior to modify the rules or allow exceptions. Proposed changes to the executive order banning poisons would be interesting enough for the woolgrowers—the ban, after all, struck at their pocketbooks and their way of life—but environmental and humane societies would find it hard to get their members excited about petitions requesting permission for the emergency use of Compound 1080 in certain watersheds in Montana or for experiments with licking posts and "single-lethal dose" baits. The woolgrowers counted, too, on their own plight. Who could defend the coyotes, which were in no danger of extinction, at the expense of decent people trying to make a living?

Sympathetic congressmen rallied to the cause. In 1973 the House held two hearings, the Senate one, on predator control policy and poisoning. The ranchers, represented by the National Woolgrowers' Association, told legislators that a vigorous system of predator control was essential. Without it coyotes would drive even good managers out of business. The proposition, though, required better empirical support than the sheepmen could easily muster. Estimates "of damages done or savings effected can only be the wildest guess work," Gabrielson had warned his colleagues in 1928. Figures like that would "ultimately do enough harm to the Bureau to more than overbalance any temporary notoriety and assistance they may occasion."[3] Now, the harm was becoming apparent. Environmental and humane societies were openly skeptical of the ranchers' claims and of estimates made "to satisfy Congress" during the days when there had been general acceptance of poisoning.

Opponents did not have to fight hard on this issue; there was not even agreement on what the figures were, much less what they meant. Nathaniel P. Reed, Assistant Secretary of the Interior for Fish and Wildlife and Parks, told the House Agriculture Committee that no one knew the extent of predator damage. The Fish and Wildlife Service, he said, had never attempted to keep such records; it would be too expensive. One of the main goals of the new research program was to develop more accurate methods of finding out how much stock predators took. What studies there were, he added, did not support the ranchers' claims. They showed that losses to predators were low—less than 3 percent of the flocks. Robert W. Long, Assistant Secretary of Agriculture, followed Reed to the stand. He told the committee that losses to predators were $50 million a year and were the limiting factor in the survival of the sheep industry. Predation losses had risen from 4.1 percent in 1950–1954 to 8.1 percent in 1970 and 7.2 percent in 1972. Questioned, he did not disagree with Reed that the data were not as good as they could be but said they were the best available (he did not say how good that was). The committee turned back to Reed, who said that losses of sheep even during years of major poisoning were around 7 percent.[4]

Woolgrowers had relied on Congressional inertia—the tendency to preserve programs that were in existence. Now they found this inertia working against them. Opponents of poisoning pointed to conflicting statements, poor data, and biases in the woolgrowers' statistics. They urged waiting. They suggested that new studies would provide a better basis for action. Representative Dingell and twenty-one co-sponsors (surely enough support) introduced a bill that would have continued the restrictions on poison but authorized more state predator control and funded research on new methods and the effectiveness of the program. Congress held more hearings. It considered this and other bills. It took no action.

The woolgrowers had more success with the executive branch. President Nixon did not help them—possibly due to the press of other business—but Vice President Gerald Ford did. In July 1975 he modified Nixon's executive order to allow "experimental" use of the M-44, a descendant of the old "coyote getter." Almost the only difference between it and the original was that the newer device used a spring instead of a small explosive to fire a dose of cyanide into an animal's mouth when it took a bait.[5] The "experiment" seemed better designed to test public reaction to poisoning and the woolgrowers' reactions at the polls than to test the effectiveness of the device against coyotes.

Organizations like the Fund for Animals and Defenders of Wildlife were outraged. Groups less committed to humane action had a different response. Thomas Kimball, executive vice president of the National Wildlife Federation, had written to express the NWF's opposition when Ford first considered revoking the 1972 ban. The program, he said, was not cost-effective and killed many nontarget animals. When Ford allowed the use of coyote getters, Kimball wrote again—to express satisfaction that the substance of the 1972 order had been retained. The administration, he realized, had to do something for the ranchers (particularly since Ford would run for president next year). From the environmentalists' point of view, the M-44 was the least of the available evils. It was single-shot, posed no danger of secondary poisoning, and when properly used was less dangerous to nontarget wildlife than other forms of poisoning.[6]

Caught in the middle, the Fish and Wildlife Service increased its research on alternatives to 1080 and its support for outside work, concentrating on selective or nonlethal methods. Coyote-proof fencing had been a dream since the early part of the century, but work was renewed. Sheep dogs, casualties of the drive to reduce costs, enjoyed a comeback. People began experiments (still continuing) with guard dogs—animals that would not herd the sheep but live with the flock and protect it.[7]

There were more exotic methods. One was a 1080-filled sheep collar. A coyote, leaping for the sheep's throat, would puncture the collar and poison himself. In the early 1970s the collar was an argument against the bait stations, and it may have been used to convince President Nixon that he could ban poison baits without incurring the ranchers' wrath. After the ban promising test results helped keep poison out. Help, it appeared, was just around the corner. Experiments continued into 1984 and they may be revived. Then there was "aversion therapy." Pieces of wool or mutton soaked in chemicals that made coyotes sick were scattered over the range. In theory, coyotes would eat the baits, come to associate sheep with vomiting, and steer clear of the flocks. This did not prove completely effective—nor did attaching capsules of synthetic skunk odor to the prey's ears or doping baits with birth-control chemicals.[8]

Even as research went on, the political situation that had produced the poisoning ban was changing. By the mid-1970s the bipartisan environmental coalition that had passed the endangered species acts and banned Compound 1080 was dissolving, a casualty of victory and the usual exhaustion that overtakes political crusades. The troops dispersed, leaving to the bureaucrats, cabinet officers, and environmental lobbyists the dull but essential work of writing and enforcing rules. Now the personal commitment of the president and his appointees became more important. The recent history of the predator control program shows the differences between the Carter and Reagan administrations on environmental policy.

President Carter believed in environmental action and his administration's predator control policy reflected this view. In May 1977 he sent a message to Congress setting out guidelines for the Animal Damage Control (ADC) program, as it was now called. These were the same guidelines Nixon had announced in 1972. There was to be no poison and no general reduction of predator populations. Coyote-sheep contact would be minimized by control measures. Where killing was necessary, only the offending individual animal would be removed. Secretary of the Interior Cecil Andrus reinforced this program with his own directive. The reaffirmation of the Nixon policy had the unintended effect of uniting the woolgrowers, the Department of the Interior, and most of the anti-poisoning forces behind a full review of the program. The Department needed evidence that its policies would do the job; the woolgrowers were sure a review would show that poison was needed, and the environmental and humane groups thought it would prove the opposite.

In January 1978 Andrus responded to the woolgrowers by forming an Animal Damage Control Policy Study Advisory Committee. It was as unwieldy as its name. Composed of representatives of conservation groups, stockraisers, and interested agencies, it was to "summarize all pertinent information" on the current program, control methods, and the western sheep industry. It held hearings in Washington and in the West and met three times. Its report presented the alternatives, assessed the data, and said what had been done.[9] This left Andrus free to do what he wanted.

Meanwhile, Interior's own Office of Audit and Investigation reviewed the program. Its report echoed, with more statistics, the indictment made by the Leopold and Cain committees. The accounting system, it said, did not provide meaningful financial data. The use of different forms made it impossible to compare expenses from one state to the next, and to determine what stock was protected, what was lost to predators, what numbers were causing damage, or what was the relation of control work to any reduction in predator damage. Financial records were incomplete. As much as $3.4 million from "direct cooperators" (livestock associations and states paying for control operations) could not be accounted for. Managers did not adequately monitor results or conduct evaluations of individual programs. Funds were allocated by "historical funding patterns rather than being justified by a thoroughly documented analysis." Program officers used outdated agreements. The policy handbook had not been revised even to take account of the 1972 ban on poisons. The program, the Office concluded, was poorly administered and could not be shown to be of value.[10]

Outside the government, pressure was mounting for further change. In 1971 the Fish and Wildlife Service had convinced Defenders of Wildlife to drop its demand for an environmental impact statement covering the predator control program by pledging immediate action on the poisoning issue. Nixon's ban had fulfilled that promise, but it had only bought the agency some time. By the mid-1970s opponents were again demanding that the Service "comply" with NEPA. It did. In 1978 it circulated "for comment" a Draft Environmental Impact Statement.[11] It was, even for a government document, vague. It described the current program and proposed several alternatives, but it had no clear direction.

Reaction was swift and negative. The USDA's Office of Environmental Quality Activities said that "The Draft Environmental Impact Statement fails as a quality document and we recommend it be prepared in accordance with the Council on Environmental Quality (CEQ) regulations."[12] The Bureau of Land Management suggested that the draft take into account two major recent policy statements, the president's message of 23 May 1977 and the Secretary of the Interior's directive of 31 May 1977. It found deficiencies "so numerous and major that the draft clearly is not adequately responsive to the impacts, issues, and established environmental assessment requirements."[13] The Council on Environmental Quality suggested an entirely new draft. This one "[i]n essence, breaks away from the Administration's stated policy on predator control."[14]

Reactions from outside the government were less restrained. John Grandy of Defenders of Wildlife called the statement "inaccurate, occasionally incomprehensible, often dishonest, and totally inadequate."[15] Elvis Stahr of Audubon and Thomas Kimball of NWF wrote a joint letter asking that the draft be redone. It could not be used to make decisions about the program and therefore did not comply with the intent of the law. In discussing losses to predators it contradicted the Office of Audit and Investigation's statement that the statistics collected did not allow a measure of the resources protected or the effectiveness of the program in abating damage. There had not been adequate provision for public input.[16] Letters from individuals and other organizations echoed these themes.

Interior insisted that the draft did discuss the president's policy, did deal with nonlethal alternatives, and did concentrate control efforts on the individual offending animals. It made some minor modifications and printed all the responses to the first draft as part of the final statement.[17] It was still impossibly vague, but that was probably deliberate. Interior was under pressure from the environmental movement and humane societies but also from Westerners already upset by Carter's policies on irrigation and western land management. The statement avoided choices and satisfied no one, but it had the virtue of making the Department no new enemies.

The difference between this process and the one that had brought and then dealt with the Leopold report some fifteen years earlier showed what had changed—and what had not. In 1962 there has been no formal way for opponents of poisoning or PARC policies to make their voices heard. They had had to rely on a sympathetic Secretary of the Interior and the political weight of organizations that supported parks and wildlife programs. Now they could force a declaration of policy; the Secretary had to take account of "sensitive problems and divergent viewpoints."[18] Predator control had been part of "wildlife" policy. A few groups and a limited part of the public had been concerned. Now the issue was part of "environmental" policy. Many more people were interested—as witness reaction to the draft statement. On the other hand, the ranchers still dominated the actual conduct of the program (the Office of Investigation and Audit's report made that clear). Despite the strength of environmental groups, the Department of the Interior could not, even with a sympathetic president, end the poisoning.

In January 1980 Andrus summarized recent policy changes for an Animal Damage Control Conference in Austin, Texas. Research on 1080 was being ended. The poison "will never be accepted by a majority of our society in any form." Denning—digging out coyote dens and clubbing the pups to death—was also being stopped; too many people found it "repulsive and inhumane." The Department would try to reduce predator-livestock conflict and conduct control operations against offending animals only. It would also help farmers who suffered losses. No one, he promised, would have to go out of business because research was going on.[19]

Ronald Reagan's election threw those plans into disarray. Reagan's political base was the West; ranchers and western developers had supported him; and Reagan had no commitment to the environmental movement or its goals. His rhetoric as presidential candidate had been relentlessly upbeat and pro-development. He dismissed the idea that we were running out of resources or destroying nature. His Secretary of the Interior, James Watt, was even less committed to environmental protection. He had made his reputation working for the Mountain States Legal Foundation, opposing environmental legislation and environmental lawsuits. Everyone expected he would continue doing so as Secretary of the Interior, and he did not disappoint them.

In September 1981 Watt directed the Fish and Wildlife Service to explore "all management alternatives for the Animal Damage Control program." He had, he said, reviewed previous policies and administrative records and found that "past Secretarial decisions have not always been based on the best available biological information" (he did not elaborate on that statement).[20] Two months later Robert Jantzen, Acting Director of the Fish and Wildlife Service, wrote to the agency's regional directors, enclosing a copy of Watt's letter. It was time, he said, to resume research on 1080, "including, but not limited to, its use in single lethal dose baits or other selective techniques for selective delivery." He reversed Andrus's policy on denning. Where coyotes were killing stock to feed their young, the cubs could be killed, if it was done humanely (by gassing or shooting). Early in December the Service filed notice of its intent to challenge the order canceling and suspending the use of 1080. It said that the evidence supporting the ban was less than adequate and that research since 1972 suggested that 1080 could be safely used. The Service would consider applications to use the chemical in single-dose lethal baits, lethal collars, and bait stations.[21]

A year later an EPA administrative law judge, Spencer Nissen, ruled on the various motions. He allowed the poison collars and, with restrictions, single-dose baits. He denied all other applications. Predictably, this satisfied no one. Defenders of Wildlife and twelve other organizations filed an exception, asking that no uses of 1080 be allowed.[22] Wyoming and South Dakota protested Nissen's refusal to allow bait stations. The Fish and Wildlife Service and the Department of Agriculture objected to the restrictions on single lethal dose baits. Another round of hearings produced only another EPA decision affirming Nissen. It rejected the stations, kept the restrictions on single-dose baits, and reinforced those on the collars.[23]

The legal battle still goes on, not in public view but in the subterranean world of notices in the *Federal Register*, administrative hearings, exceptions, and legal review of agency decisions. The outcome is in doubt, but the pushing and hauling of the last decade have established a rough agreement on some matters. Everyone realizes that the public will not stand for clubbing coyote pups over the head with a shovel. Large bait stations are almost as unacceptable, but selective control may pass public scrutiny. Most people accept the ranchers' need to keep their stock safe but oppose a return to the old program or oppose the use of poison in any form.[24] Killing coyotes to protect sheep is becoming one of those programs that is opposed by many and held in place by the determination of a few. Its existence is an outrage to its enemies, the limitations on it irritating to its supporters.

PROBABLY the major problem facing government agencies managing the wolf today in the United States is that of striking the right balance in preserving the species where it still exists while minimizing its conflict with men, and, in doing so, dealing with biological realities while absorbing emotional pressure from public groups that either believe that no wolf should be killed or that all should be.[25]

Wolf management is a different proposition than coyote control. Coyotes are abundant; wolves are scarce. Economics drives (at least in theory) coyote killing, but there have not been enough wolves in the United States since the 1920s to make that an issue for the larger predator. Wolves, too, arouse much more emotion. Coyotes are, at most, a romantic symbol of the West (usually howling in the distance as you sit around the campfire). Wolves are either horrible creatures who must be exterminated or resident spirits of the wilderness who must be preserved. Finally, coyote "management" was an ongoing project that had to be modified to take account of new public attitudes. Wolf programs had to be built as the public enthusiasm for wilderness, wildlife, and endangered species made preservation desirable.

The building of a wolf program was a rough process, particularly when public enthusiasm ran counter to professional judgments. The distinction between the two was most apparent in Alaska, where managers clashed with a public that had learned the "lessons" of ecology. During the 1960s the Alaska Department of Fish and Game and the U.S. Fish and Wildlife Service had worked out management plans that allowed research, minimized wolf killing, and satisfied hunters. That delicate balance was lost when conditions changed. "The 1970s," said one summary of Alaskan wolf management, "were characterized by sharp increases in wolf numbers, declining ungulate populations, state-initiated control operations, and intensely complex litigation."[26] The first two caused the third, which in turn brought on the fourth. All of it depended on a factor the authors of the summary did not analyze: the rise of environmental and humane organizations determined to protect all the wolves—each and every one.

As populations of deer, moose, and caribou fell and wolf numbers rose, Alaska Fish and Game officials began to think about wolf control. They did not argue, as had been done in the 1920s, that control was needed to assure a supply of game for hunters. They believed that natural factors—weather, disease, or lack of food—had reduced the numbers of several species. The wolves, having turned to other food, had declined along with their primary prey. Now, Alaskan wolf biologists believed, the wolves might keep populations of the slow-breeding prey species down for many years to come. If some wolves were killed now the prey population would recover much faster. Their recommendations, made for a particular situation and based on recent research, clashed with the now-conventional wisdom. The wolf-loving public assumed a balance, believed that wolves and their prey were dependent on each other, and thought the system would automatically stabilize with all species at a high level. Anti-wolf factions, on the other hand, saw the situation as confirmation of their beliefs that wolves could not be let alone or they would wipe out the game. Neither side easily accepted the complex picture the wolf biologists now held.

In 1975, when the state began an aerial gunning plan, the Fairbanks Environmental Center, Friends of the Earth, and private individuals quickly brought suit to stop it. They were successful, and the Alaska Fish and Game Department tried to get around the problem by asking the National Association of Audubon Societies to study its wolf management operations. That fell through when it became apparent that the Department wanted to continue killing wolves while Audubon investigated. The Society withdrew, saying that its reputation would suffer if it were thought to be aiding a program of wolf killing.

Defenders of Wildlife went back to court in January 1976, challenging the state's revised plans. Operations were to include some federal land near Fairbanks, and Defenders asked that an environmental impact statement be prepared before the program went forward. The organization also sought Congressional action. In 1976 four bills were introduced to end aerial gunning and large-scale killing. They failed to pass, in part because of objections raised by the U.S. Fish and Wildlife Service. The court cases, with new motions, appeals, and hearings, dragged on into 1980 and involved practically every federal-state issue of interest to Alaska. Only recently has wolf control shown signs of "becoming more of a routine management activity and less of a special, high visibility event requiring extensive public hearing and debate."[27]

Until the passage of the Endangered Species Act of 1973, the Fish and Wildlife Service had little responsibility for wolf management in the "lower forty-eight." Then it faced three different situations, each with its own version of the "major problem" that Mech and Rausch referred to: "striking the right balance." There was the small but viable wolf population in northern Minnesota. It had to be managed, with particular attention given to the problems of hunters, farmers, and wolf preservationists. The northern Rockies contained wolf habitat and, reports suggested, stragglers, but there were no packs. There was sentiment for reintroduction (Glacier and Yellowstone National Parks were the favored spots), but ranchers opposed it. The third problem was the red wolf. Biologists believed the hybrids would soon swamp the pure-bred population. Unless there was some immediate action the species would vanish.

Public, scientific, and bureaucratic attention focused on the Minnesota wolves. There was strong anti-wolf feeling among the people of the area, but immediate problems with the Endangered Species Act itself. The law aimed to restore species to their place as part of a functioning ecosystem. Wolves, though, had ranged over the entire continent, on land now given over to farms, towns, cities, and suburbs. How much would become wolf country again? What interests would yield to the wolf's and who would decide? The Eastern Timber Wolf Recovery Team had to reconcile Americans' dreams of wilderness purity with the realities of human occupation of the continent and the still widespread fear and hatred of wolves.

The first problem was to decide what the recovery team was to do. Though the Fish and Wildlife Service had been helping species recover since the 1930s, the Endangered Species Act called for a much broader program. The wolf team was one of the first appointed, and the Secretary of the Interior took a conservative approach. He chose a team from the organizations that had or would have responsibility for managing the wolf. There were three state representatives, one each from the Michigan, Wisconsin, and Minnesota Departments of Natural Resources. Two members came from the U.S. Forest Service, one from the Park Service, and two from the Fish and Wildlife Service (one was wolf ecologist L. David Mech).

The group decided that its job was to produce "a purely biologically based plan." It would "disregard possible political or social considerations" and write an "ecologically sound plan for the maintenance, enhancement and recovery of this subspecies throughout as much of its present and former range as feasible."[28] Despite this brave declaration the team necessarily considered political and social factors. The wolves' survival, it declared, depended on prey, vacant land, ecologically based management plans, and sufficient knowledge to make plans. All but the first involved humans (on Isle Royale humans could be controlled, but not on the mainland). The team included in its "ecologically sound management" plans not only protection for wolves but "regulated taking" in marginal areas of the range. This would minimize stock losses and anti-wolf sentiment. It insisted too on "concerted efforts at public information and education" as part of the recovery plan.[29]

The team met the problems of human-wolf interactions by creating a complex system of zones where wolves would be managed by different plans for different objectives. There was to be a sanctuary, Zone 1, in the northeastern part of the state, centered on the Boundary Waters Canoe Area and the Superior National Forest. Here wolves would be allowed to develop a natural social structure under natural conditions. In a buffer zone around that area (Zone 2) and on a tract to the west (Zone 3) wolves would be kept to one per ten square miles. A wider area surrounding all these (Zone 4) took in the northern third of the state. Here wolves would have a population of one per fifty square miles. A kill of one hundred wolves per year would be allowed in that area. Another sixty wolves, the team estimated, would have to be killed for damage control and sixty more would be killed illegally. Killing might be necessary, but it was illegal, since the population was classified as "endangered." The team had to propose—as part of its plan to save the animals—that the Minnesota population be reclassified as "threatened."

In December 1975 the wolf management team sent out a draft plan for criticism. Three dozen copies went to people in state and federal agencies, twenty-seven to conservation, hunting, and preservation groups, and a dozen to academic biologists. Reactions were mixed. Other federal agencies were most concerned that the team had not taken their needs into account. The Bureau of Indian Affairs pointed out that the Red Lake Indian Reservation lay in Zone 3 and that the Indians were not entirely happy with the prospect of wolves killing "their" deer. Nor did they particularly like the plans to manage the wolves by managing the deer. Since the Indians were not subject to Minnesota hunting laws, wolf management would require some unusual bargaining. The Forest Service protested that the recovery team had ignored multiple-use management goals. Selective timber cutting to raise more deer to feed more wolves was not a good timber management program. The team also had not considered the public. It had not asked for outside comments or even allowed people to express their wishes. The Service wanted major revision in format, recommended actions, and in the review process before the plan went into effect.

The most negative reactions came from Minnesota. The Minnesota member of the recovery team, Leroy Rutske, noted that the state Department of Natural Resources had asked that the Minnesota wolves be "delisted" (it would renew that request when the draft was sent out). The state could manage its wolves and should be allowed to do so. "Evidence presented during deliberations of the Team has not weakened Minnesota's position but has strengthened it." He opposed the sanctuaries. Animals dispersing from there would "increase local public hostility toward wolf management programs."[30]

That was already high. The Coleraine Civic Club stated that its members did not think wolves were either endangered or threatened. They saw no need for the proposed recovery plan or for wolf research. The Virginia (Minnesota) Sportsmen's Club protested that there had been only one Minnesota man on the team, whereas there were "five federal bureaucrats." The plan was an attempt "to insure job stability for those involved, . . . a back door approach to people control and gun control." They too wanted research funds cut off. "These repetitive studies serve no valid purpose, as the necessary facts have been known since the turn of the century."

Wildlife protection groups split, usually over plans to kill some wolves to save the rest. The National Wildlife Federation, the Northern Midwest Region of the National Association of Audubon Societies, and even HOWL thought the plan was generally sound. The Committee for Humane Legislation was "shocked and amazed" that the plan allowed the taking of an endangered species. It suggested "keeping man out of the wolves' territory rather than claiming the land as man's territory and moving the wolves out." Defenders of Wildlife, joined by nine other organizations, vigorously opposed killing any wolves. John Grandy, Defenders' executive vice president, said that taking should be restricted to extraordinary problems. People grazing their stock on public land must accept losses as part of the risk of using this area. As for the team's actually planning for an illegal kill, that was "ridiculous." Better management and stronger law enforcement was the answer.[31]

The director of the Forest Service's North Central Forest Experiment Station, Jack Ohman, raised more basic questions. What, he asked, was "critical habitat"? The team had not indicated any "critical" areas outside Minnesota. "[A]re there no other critical habitats?" Even within the state the team was not following the law. Was it not, he asked, "really designating critical and sort of 'semi-critical' habitat areas in Minnesota? Should not the concept apply to the full former range (or present potential range?) of the wolf? Also, if the full intent of the act (return to and/or maintenance of viable population levels) applied outside of critical habitat areas, then the team's recommendation for reduced population levels based on social considerations was not valid (assuming that this population level is not a 'viable' level)."[32]

Ohman explicitly raised the issue of how much land could, or would, be given over to the wolves. Reaction from states that would host the reintroduced animals suggested that there was little available. The recovery team had identified eight places as potential wolf range—northern Wisconsin, the Upper Peninsula of Michigan, and six areas in the east, ranging through the Appalachians from Maine to Georgia. Maine's Commissioner of Inland Fisheries and Game was upset that "without prior consultation" the team had raised the issue of reintroduction and even assigned the state a role. Maine had too little money to undertake such a program, and local attitudes, he warned, made wolf introduction "socially unacceptable." New Hampshire, Georgia, North Carolina, Wisconsin, and Michigan reacted in much the same way. In varying degrees, they warned that most people in the area would be hostile to the program.

The concern for local attitudes that ran through the recovery team's reports, its allowance for an illegal kill, and the comments from game officers across the country were a warning. Pro-wolf sentiment was strongest where the animal was not and never would be—in settled areas. It was weakest in the backwoods and primitive areas where the wolf would live. This was more than an academic concern. When the Fish and Wildlife Service had released four radio-collared wolves on the Upper Peninsula of Michigan in 1974, all four had been killed. One had been left on a game warden's doorstep.[33] This was repeated in other areas. All the recovery teams faced major problems convincing people they could live with wolves.

But the people who had pushed through the endangered species act wanted it enforced. State and federal authorities heard from HOWL, FATE (Fur and Trapping Ethics, another Minnesota group), Canadian and American Wolf Defenders, the Texas Council for Wildlife Protection ("Dedicated to adding Compassion to Conservation"), the North American Wolf Society, and many private individuals. Caught in the middle, the Fish and Wildlife Service tried to make management plans that would satisfy everyone. In 1978 it got the wolves of the continental United States, regardless of subspecies, classified as "endangered," with the exception of the "threatened" Minnesota population. The change allowed individual wolves of the "threatened" population to be killed—to save the rest—and it protected, at least on paper, the stragglers in the Rockies.[34]

The main problem facing the Northern Rocky Mountain Wolf Recovery Team was the intense local hostility to the reintroduction of wolves. Ranchers around Yellowstone National Park were particularly adamant. They had complained in the 1930s when the Park Service had ended coyote control; they were concerned about grizzlies from the park; now the Fish and Wildlife Service proposed bringing back another animal to eat their stock. The team also had little information about how wolves had used the area. Ranchers and trappers from the Biological Survey had killed the last breed in the Rockies some fifty years earlier, so there was no counterpart to the research program that had been going on on Isle Royale since 1957 and in Minnesota since 1968. Possibly because of these problems, the Rocky Mountain Wolf Recovery Team included representatives of the Fish and Wildlife Service, the Forest Service, the National Park Service, the Bureau of Land Management, the Montana and Idaho wildlife commissions, and the National Audubon Society, as well as a wolf biologist from the University of Montana.

The team produced an "in-house" draft of a recovery plan in 1976, but two years passed before it sent out a draft for criticism. The final plan was not ready until May 1980.[35] The plan was much less detailed than the one for the Eastern Timber Wolf and the team was much more hesitant about its final objective. It was pessimistic about the chances of a successful reintroduction. There were, it said, few places where wolves could be reintroduced—and it did not identify them. Though the establishment of two independent populations was the ultimate goal, the immediate objective was the determination of areas where work could start. There would have to be an enormous amount of biological work, public education, and preparation even for this to be done. The team did not plan even to "determine feasibility of re-establishment" (as the flow chart put it) until 1985, and it set 1987 as the date for actually bringing in wolves.[36]

The red wolf was yet another story. Secretive, shy, without call or legend, it was the forgotten canid. That biologists developed a program to save it—one that required considerable work and money—is more a testament to their interest and the public enthusiasm for wild-life preservation than to any sentiment to preserve this species. The work began quietly and has proceeded without the blaze of publicity that has accompanied work in Minnesota. The Fish and Wildlife Service put the red wolf on its first endangered species list and in 1967 began considering plans for recovery or at least rescue. It selected the species for priority treatment under the 1973 Endangered Species Act. By November of that year it had signed an agreement with the Point Defiance Zoo in Tacoma, Washington; the agency was to supply animals, the zoo to breed them. If all else failed there would at least be some in captivity. With the cooperation of the Louisiana Wildlife and Fisheries Commission and the Texas Parks and Wildlife Department, the Service began an expanded program of research on the wild populations.[37]

By 1975 it was clear that the species could not be saved in the wild. There were too few pure red wolves left and their habitat was too small and disturbed. The hybrid swarm would, within a few years, swamp them. The Service shifted its efforts to saving the species in captivity until it could be reestablished somewhere else. Biologists realized that live-trapping would hasten the end of the red wolf in the wild, but that was unavoidable. They began live-trapping everything in the Gulf Coast swamps where the last wolves remained. Each specimen was examined and those that appeared to be really red wolves were sent to Point Defiance. By 1980 the free-ranging wolves were gone. There were in captivity a dozen wild-caught, another dozen captive-born, and three dozen wild ones that might or might not be the genuine article. Research in several areas was going forward, and the team was making plans to reintroduce the red wolves into other parts of their historic range by the mid-1980s.

The plans, programs, and hopes of the wolf recovery teams, like the struggle over predator poisoning, indicate how far Americans had come toward accepting these predators as part of the landscape—and what distance remained. Saving the wolves was important to many Americans, but there were significant numbers, particularly among those who lived in wolf country, who wanted the creatures dead. In the last few years, however, the situation has been changing. We now seem to be working out a place for the wolf and the coyote in our civilization, coming to some terms—for this generation at least—with nature.

Epilogue

The wolf poses one of the most important conservation questions of our time. Will the species still exist when the twentieth century passes into history? Or will man have exterminated the wolf as a final demonstration of his "conquest" of the wilderness and of wild things that dare to compete or conflict with him?[1]

In 1973, twelve years after Canadian biologist Douglas Pimlott asked these questions, his colleague, John B. Theberge, could say that "an answer seems to be emerging. Yes, the wolf will still exist."[2] Today the answer looks clearer. The Minnesota wolves range their forests. A Canadian pack, with no regard for the Northern Rocky Mountain Wolf Recovery Team, crossed into Glacier National Park in 1985. The next spring the wolves produced the first documented wolf den in the western United States in fifty years. Six months after this the Fish and Wildlife Service shipped eight red wolves to North Carolina to be acclimated to their new home in the Alligator River National Wildlife Refuge. A pair pupped in the spring of 1987. If everything has gone well, the red wolves will be roaming Dare County as you read this. There seems to be a place in modern America for the "beast of waste and desolation."[3]

It is not just Americans who are keeping the wolf alive. The environmental movement that swept over the United States had counterparts abroad, and wolf protection is an international effort. In 1973 wolf biologists formed the Wolf Specialist Group of the Species Survival Commission of the International Union for the Conservation of Nature and Natural Resources. At their first meeting they produced a "Manifesto and Guidelines on Wolf Conservation." Wolves, the document said, "like all other wildlife, have a right to exist in a wild state. This right is in no way related to their known value to mankind. Instead it derives from the right of all living creatures to co-exist with man as part of natural ecosystems." The scientists called for preservation of local wolf populations under conditions that preserved the social structure of the pack. They recommended research, public education, and international cooperation.[4]

173

A symposium in 1979 heard fifty-one papers by researchers from around the world. By 1980 the Wolf Specialist Group had members from twelve countries and was seeking representatives from other countries with wolf populations. Norway, Sweden, and Finland began a cooperative effort to save their wolves. Poland started a program to restore wolves in a forest preserve near the Russian border. In Russia scientists increased their efforts to manage and preserve several local populations.[5] Italian scientists, with the aid of David Mech, began trapping and radio-collaring wolves. At present (1987) there are about 150 wolves in Italy, some only twenty to thirty miles from Rome.[6] For the world as well as the United States a new equilibrium, a new relationship between humans and wolves, seems in the making.

Americans are also working out a new relationship with the coyote, though the circumstances are somewhat different. The "brush wolf" has flourished and spread under persecution. It is now established in much of the northeastern United States, and individuals have been reported, and shot, as far south as Virginia and West Virginia.[7] In the West the coyote still roams the range, but it has also found a new home. As sprawling western cities expanded into the desert, coyotes began to raid garbage cans. They turned to pets; people began to report finding their cats on the front lawn, eviscerated. Three coyotes were seen pulling down a poodle on a traffic island in the middle of a busy street—in daylight.[8] There have been reports of coyotes attacking children, and in 1981 it was alleged that a coyote killed a three-year old in Glendale, a Los Angeles suburb. There were newspaper stories and demands for control or extermination of coyotes in the area. Death threats were reported against people who spoke up for the wildlife.[9]

BUT THE FATE of wild canids in North America depends more on human attitudes than on their adaptability. History counts for more than natural history. Central to that history is our continuing search for a relationship to nature that is suitable to our knowledge and the conditions of our lives. We need a myth for nature in industrial America. In our search we have poured old wine into new bottles. All the important themes of modern environmental nature preaching—nature as a refuge, humans' need for harmony with the world, the concept of the world as a full, ordered and beautiful place—are deeply rooted in Western civilization. Wilderness was a refuge for the Israelites escaping from Egypt and for the prophets of the Old Testament. It was the place where Christ prepared for his public ministry (and where he met the Devil). The Greeks debated whether there was a natural order and what it was. They introduced the speculation that the world is filled with every conceivable organism and that nature has a place and a purpose for it. Every generation has built and rebuilt on these foundations.[10] It is just that we couch our discussions in terms of ecosystems and predator-prey interactions, where our ancestors spoke of animating spirits or God's personal intervention.

This is not to say that ecology begins with Aristotle or even with the Scholastics—only that we have interpreted science in terms drawn from Western civilization and American experience. Our culture does not, despite the popular view, see nature only as something to be destroyed. "Conquest" and "citizenship in a biological community" are the ends of a spectrum over which our culture has ranged. That we must dominate nature has been the ruling idea for centuries, but there have always been dissenters and a dissenting tradition. Now this dissent has become visible as we became less fearful of nature and less subject to its harsher realities. Science has made respectable ideas that have always been there but have only recently become popular.

We have worked in familiar ways just as we have thought in familiar terms. Environmental politics has relied on the same mechanisms used by other causes: voluntary organizations—public, private, or professional—it has worked through affected agencies, interested legislators and lobbyists. It has formed ties to agencies, and used public propaganda. The movement has borrowed from civil rights and the counterculture, and it used litigation (for a time) more than most groups; but even that has yielded to the same realities that keep many disputes out of court. Lawsuits are expensive, time-consuming, and rarely give satisfactory results. All but the most extreme groups have come to participate more and more in the process of cooperation and compromise by which agencies form policy from laws. In May 1987, for example, Defenders of Wildlife sponsored a day-long symposium on wolves. The director of the Fish and Wildlife Service spoke and a Montana sheep rancher opposed to restrictions on predator control participated in a panel discussion. Defenders has, in fact, been working with ranchers for the last two years, attempting to find common ground and an acceptable program that will preserve wildlife and sheep.[11]

The role and impact of "science" in the transformation of American ideas is ambiguous at best. Science is a body of knowledge and a source of authority to which people appealed, a myth, and a social activity. While it is supposed to guide wildlife policy, that activity unfortunately involves choices and values more than decisions of fact. Scientific ideas can, in any case, be interpreted in a variety of ways. There is also a wide spread in our understanding, ranging from that possessed by the professional biologist to the knowledge of the "educated layman," whoever that person is. Nor can we isolate science. It began to describe the natural world in detail at the time the last large natural areas in North America began to disappear. Our enthusiasm for nature has been nurtured not only by the scientists' vision but by the knowledge that the wilderness and its wildlife were vanishing and by our own increasingly urban way of life.

It is possible to say that science, as knowledge and mythic authority, has played a key role in the shifts in ideas and policy of the last century. Darwinian evolution forced a reevaluation of our life in nature and set limits to what the "educated" could accept as a true picture. Ecology gave us a detailed knowledge of how the natural world functioned. Science has overturned some ideas and made others more plausible or attractive. Scientists themselves have played important roles in the public debate over "man and nature." They propounded and explained the new theories. Often working through their professional associations, they led protests against the destruction of wildlife and wild lands. They were prominent among those fostering an appreciation of wilderness and wildlife. The two major nature writers of this generation, Rachel Carson and Aldo Leopold, were both scientists, both deeply interested in ecological ideas. Even Farley Mowat's *Never Cry Wolf*, a voyage of self-discovery, is cast in the form of a scientific investigation.

That science, a complex result of a sophisticated culture, guides our appreciation of the primitive is ironic but not surprising. Americans are awed by science, have a passion for gadgets, and are deeply nostalgic about the frontier past. We have as a society often turned to modern means to preserve old virtues and values. The wolf with the radio collar, providing data for scientists to use in reestablishing the primitive ecosystems of North America, may be the perfect symbol of our efforts to come to terms with our knowledge of nature's order, our power over it, and our need to preserve our mythic past.

Notes

Preface

1. Henry Nash Smith, *Virgin Land: The American West as Symbol and Myth* (Cambridge, Mass.: Harvard University Press, 1950), v.

2. On attitudes toward nature, one should start with Clarence Glacken, *Traces on the Rhodian Shore* (Berkeley: University of California Press, 1967), a magisterial treatment from antiquity to the eighteenth century. Keith Thomas, *Man and the Natural World* (New York: Pantheon, 1983), goes from the fifteenth century to the present. Roderick Nash, *Wilderness and the American Mind*, 3rd edition (New Haven: Yale University Press, 1983), and Alfred Runte, *National Parks: The American Experience* (Lincoln, Nebraska: University of Nebraska Press, 1979), cover the American experience.

3. William Hornaday, *The American Natural History* (New York: Charles Scribner's Sons, 1904), 22, provides an excellent example of even "enlightened" opinion. See also Theodore Roosevelt, *The Wilderness Hunter*, Volume II of *The Works of Theodore Roosevelt*, National Edition (New York: Charles Scribner's Sons, 1926), 305.

4. Hornaday, *American Natural History*, 22; Roger Caras, *The Custer Wolf* (Boston: Little, Brown, 1966), 25, 30, 59. New ideas about nature and wolves have reversed even the werewolf story; see Ursula LeGuin, "The Wife's Story," *The Compass Rose* (New York: Harper and Row, 1982), 221–225. The "good wolf" may be seen in Farley Mowat, *Never Cry Wolf* (Boston: Little, Brown, 1963); Barry Holstun Lopez, *Of Wolves and Men* (New York: Charles Scribner's Sons, 1978); R. D. Lawrence, *In Praise of Wolves* (New York: Henry Holt, 1986).

5. Hope Ryden, *God's Dog* (New York: Viking, 1979), and J. Frank Dobie, *The Voice of the Coyote* (Boston: Little, Brown, 1949), show changing sentiment. A more extensive history, particularly of predator control, may be found in Stanley Paul Young and H.H.T. Jackson, *The Clever Coyote* (Washington: Wildlife Management Institute, 1951), contemporary pro-control sentiment in Charles L. Cadieux, *Coyotes: Predators and Survivors* (Washington: Stone Wall Press, 1983).

6. A good study in this area is Ronald Tobey, *Saving the Prairies* (Berkeley: University of California Press, 1981).

Chapter One. Saving Animals

1. Remarks of an egret hunter, about 1890, in T. Gilbert Pearson, *Adventures in Bird Protection* (New York: D. Appleton-Century, 1937), 51.

2. T. S. Palmer, "Extermination of Noxious Animals by Bounties," in U.S. Department of Agriculture, *Yearbook of the United States Department of Agriculture, 1896* (Washington: Government Printing Office, 1897), 55–68. As late as 1973, when the wolf had been virtually extinct in the coterminous United States for fifty years, twenty states still had bounties on the species. See Elmer W. Shaw, "An Analysis of Laws Related to the Bounty on Wolves in the United States," Library of Congress,

Legislative Reference Service, 5 February 1970. Copy from Wolf files, Office of Endangered Species, U.S. Department of the Interior.

3. Stanley Paul Young and Edward A. Goldman, *The Wolves of North America* (Washington: American Wildlife Institute 1944; reprinted New York: Dover, 1964), 324–335.

4. Young and Goldman, *Wolves of North America*, 335, 286–303, 337–368; Palmer, "Extermination."

5. On the garden mythology in America, see Leo Marx, *The Machine in the Garden* (Oxford: Oxford University Press, 1964), and Henry Nash Smith, *Virgin Land* (Cambridge, Mass.: Harvard University Press, 1950).

6. T. S. Palmer, "Chronology and Index of the More Important Events in American Game Protection, 1776–1911," Bureau of Biological Survey, U.S. Department of Agriculture, Bulletin 41 (Washington: Government Printing Office, 1912). The magazine of the National Audubon Society, *Bird-Lore* (to 1940, *Audubon* thereafter), provides an excellent sampling of the protection movement. On hunting, see John F. Reiger, *American Sportsmen and the Origins of Conservation* (New York: Winchester, 1975; reprinted Norman: University of Oklahoma Press, 1986); George Bird Grinnell and Charles Sheldon, *Hunting and Conservation* (New Haven: Yale University Press, 1925); and James B. Trefethen, *An American Crusade for Wildlife* (New York: Winchester, 1975). More general treatments of changes in American ideas about nature are in Roderick Nash, *Wilderness and the American Mind*, 3rd edition (New Haven: Yale University Press, 1982), and Peter J. Schmitt, *Back to Nature: The Arcadian Myth in Urban America* (New York: Oxford University Press, 1969).

7. Edmund Morris, *The Rise of Theodore Roosevelt* (New York: Ballantine, 1979), 204–225.

8. George Bird Grinnell, "American Game Protection: A Sketch," in Grinnell and Sheldon, *Hunting and Conservation*, 220; idem, "The Last of the Buffalo," *Scribner's Magazine, 12* (September 1892), 267–286.

9. William T. Hornaday, "The Extermination of the American Bison," in United States National Museum, *Report of the United States National Museum, 1887* (Washington: Government Printing Office, 1889), 465.

10. Theodore Roosevelt, *The Wilderness Hunter*, Volume II of *The Works of Theodore Roosevelt*, National Edition (New York: Charles Scribner's Sons, 1926), xxix.

11. Reiger, *American Sportsmen*, claims far too much for hunters. The early conservationists were hunters, but so were most adult men in that period, and it is hard to see, on the evidence Reiger presents, that hunting led, in quite the direct way he wishes to see, to conservation. Samuel Hays's argument in *Conservation and the Gospel of Efficiency* (Cambridge: Harvard University Press, 1959) is much more persuasive.

12. They were collected in 1856 as the *Complete Manual for Young Sportsmen* (reprinted New York: Arno, 1974).

13. Forester, *Young Sportsmen*, 32.

14. Forester, *Young Sportsmen*, 359.

15. Alan Devoe, "Robins for Sale: Five Cents," *Audubon, 49* (March-April 1947), 108–112; Thomas Nuttall, *A Manual of the Ornithology of the United States and Canada* (Cambridge, Mass.: Hilliard and Brown, 1832; reprinted New York: Arno, 1974); Mabel Osgood Wright, "What Is a Bird Sanctuary?" *Audubon, 36* (July-August 1934), 219–225.

16. Trefethen, *American Crusade for Wildlife*; Grinnell and Sheldon, *Hunting and Conservation*.

17. Frank Luther Mott, *A History of American Magazines*, Volume 3, 210–211; Vol-

ume 4, 380–381 (Cambridge: Harvard University Press, five volumes, 1930–1968). The magazines were eclectic, covering field sports, camping, bicycling, and nature study, and printing fiction and nature poetry as well. The first volume of *Forest and Stream*, for example, included articles on subjects from billiards and cricket to yachting. The military section had an article on the "dashing General Custer."

18. *Forest and Stream*, *1* (14 August 1873), 3.

19. Aldo Leopold, *Game Management* (New York: Charles Scribner's Sons, 1933), 13.

20. Sportsmen turned to the states rather than to the federal government because the states had legal control—wildlife was one of the responsibilities they assumed as newly independent colonies and reserved in the Constitution. Reservation was by implication: the tenth amendment reserved to the states powers not delegated to the United States or forbidden to the states. On this legal doctrine as applied to wildlife, see Michael Bean, *The Evolution of National Wildlife Law* (New York: Praeger, 1983), 10–17; James Tober, *Who Owns the Wildlife?: The Political Economy of Conservation in Nineteenth-Century America* (Westport, Conn.: Greenwood, 1981); Thomas A. Lund, *American Wildlife Law* (Berkeley: University of California Press, 1980).

21. *Recreation*, 1896–1905; Pearson, *Adventures in Bird Protection*, 343–344.

22. Reiger, *American Sportsmen*; Grinnell, "American Game Protection," in Grinnell and Sheldon, *Hunting and Conservation*, 201–257.

23. Palmer, "Chronology and Index"; Tober, *Who Owns the Wildlife?*

24. William Hornaday, *Wildlife Conservation in Theory and Practice* (1914; reprinted New York: Arno, 1972), 106.

25. Palmer, "Chronology and Index," notes that Maine outlawed the hunting of moose or deer by aliens (1852), New York forbade aliens to carry firearms in public places (1905), Pennsylvania stopped unnaturalized foreign-born people from hunting or owning guns (1909), and Washington required aliens to get a license and a consular certificate to own a gun (1911). This was a problem as late as the 1930s, when the Canadian government complained about Massachusetts laws forbidding aliens to hunt, fish, or trap. See Acts and Legislation, Box 3, Records of the Canadian Wildlife Service, Record Group 109, Public Archives Canada, Ottawa.

26. Roosevelt, *Wilderness Hunter*, is a good example. On Roosevelt's natural history interests, see Paul Russell Cutright, *Theodore Roosevelt: The Making of a Conservationist* (Urbana: University of Illinois Press, 1985). A contemporary view is John Burroughs, *Camping and Tramping with Roosevelt* (Boston: Houghton Mifflin, 1907; reprinted New York: Arno, 1970). Schmitt, *Back to Nature*, deals with many aspects of Progressives' ideas about nature.

27. Loren Owings, *Environmental Values, 1860–1972* (Detroit, Mich.: Gale Research Co., 1976), provides an excellent bibliography of nature writing. See also Phillip Marshall Hicks, "The Development of the Natural History Essay in American Literature" (Ph.D. dissertation, University of Pennsylvania, 1924).

28. Lisa Mighetto, "Wild Animals in American Thought and Culture, 1870s–1930s" (Ph.D. dissertation, University of Washington, 1986), is the best study of the full range of nature stories. See also her "Science, Sentiment, and Anxiety: American Nature Writing at the Turn of the Century," *Pacific Historical Review*, *54* (February 1985), 33–50. I am also indebted to Mighetto for personal communications.

29. Robin Doughty, *Feather Fashions and Bird Preservation* (Berkeley: University of California Press, 1975); Pearson, *Adventures in Bird Protection*, 207–210. *Bird-Lore* from 1899 has a fine collection of articles and photos on the feather trade and its

results. "The Aigrette," by Charles G. D. Roberts, reprinted in the collection *King of Beasts* (Toronto: Ryerson, 1967), protested against this practice.

30. While the main force of the growing humane movement was centered on the treatment of domestic stock and pets, concern for wild animals was a part of the movement. On the development of the humane movement, see James Turner, *Reckoning with the Beast* (Baltimore: Johns Hopkins University Press, 1980), Chapter 3. The English humane movement began around 1800; Americans began organizing and agitating for reform, mainly on the state and local level in the East, in the 1870s. Henry Bergh organized the American Society for the Prevention of Cruelty to Animals in 1866, patterning it after the Royal Society. The Massachusetts branch produced the first journal, *Our Dumb Animals*, in 1868. The American Humane Society was formed in 1877. On humane sentiment in the feather crusade, see Doughty, *Feather Fashions*, and Pearson, *Adventures in Bird Protection*.

31. Doughty, *Feather Fashions*; Theodore W. Cart, "The Struggle for Wildlife Protection in the United States, 1870–1900" (Ph.D. dissertation, University of North Carolina, 1971).

32. *Bird-Lore, 3* (February 1901), 40–41.

33. Mabel Osgood Wright, "The Law and the Bird," *Bird-Lore, 1* (December 1899), 203–204.

34. Mighetto, "Wildlife in American Thought and Culture," and "Science, Sentiment, and Anxiety."

35. On Audubon's raptor policy, see *Bird-Lore, 32* (November-December 1930), 446, or Mabel Osgood Wright, "Stories from a Bird Sanctuary, II," *Bird-Lore, 24* (September-October 1922), 253–255. Compare with *Bird-Lore, 32* (November-December 1930), 446; "Campaign for Hawk and Owl Protection," Resolution of the Directors, 9 May 1934, *Bird-Lore, 36* (September-October 1934), 333–335; and Warren F. Eaton, "Predators and Bird Preserves," *Bird-Lore, 37* (May-June 1935), 162–166.

36. Grinnell's comment is in "Conserve the Collector," *Science, 41* n.s. (February 1915), 229–232. It appears in a posthumous collection of essays, *Joseph Grinnell's Philosophy of Nature* (1943; reprinted Freeport, N.Y.: Books for Libraries Press, 1968), 65–72.

37. Nuttall, *Manual of Ornithology*, 4, 62, 65, 75–76, 79.

38. Alexander Wilson, *Wilson's American Ornithology* (Boston: Otis, Broaders, and Co., 1840; reprinted New York: Arno, 1974), 329.

39. Hornaday, *Wildlife Conservation in Theory and Practice*, 141–151, and *Our Vanishing Wildlife* (New York: New York Zoological Society, 1913), 80, 140.

40. *Bird-Lore* from 1899 gives many examples of this. For change in the Society's position, see Chapter 6.

41. John Godman, *American Natural History* (Philadelphia: Carey and Lea, 1826; reprinted New York: Arno, 1974), 256; Richard Harlan, *Fauna Americana* (Philadelphia: Anthony Finley, 1825; reprinted New York: Arno, 1974), 75.

42. William Hornaday, *The American Natural History* (New York: Charles Scribner's Sons, 1904), 22.

43. Roosevelt, *Wilderness Hunter*, 305.

Chapter Two. Science and the New American Nature Myth

1. The popular literature of the late nineteenth and early twentieth centuries is replete with this imagery, to the point that it is difficult to understand movements

ranging from national defense to settlement houses without reference to ideas of the "fit" and the "struggle for existence." On the dual heritage of evolution, see Donald Worster's *Nature's Economy* (San Francisco: Sierra Club Books, 1977). On Darwinism and ideas about animals, see James C. Turner, *Reckoning with the Beast* (Baltimore: Johns Hopkins University Press, 1980). Important contemporary books include George J. Romanes, *Mental Evolution in Man* (New York: D. Appleton, 1889; reprinted Washington: University Publications of America, 1975); idem, *Animal Intelligence* (New York: D. Appleton, 1883, reprinted Washington: University Publications of America, 1977); W. Lauder Lindsay, *Mind in the Lower Animals*, Volume I: *Mind in Health*; Volume II: *Mind in Disease* (New York: D. Appleton, 1880).

2. Turner, *Reckoning with the Beast*, Chapters 1 and 2. For a particular case, see Robert Darnton, *The Great Cat Massacre and Other Episodes in French Cultural History* (New York: Basic Books, 1984).

3. Jeremy Bentham, *An Introduction to the Principles of Morals and Legislation* (1789), quoted in Peter Singer, *Animal Liberation* (New York: Random House, 1975), 8.

4. Turner, *Reckoning with the Beast*, Chapter 4; Romanes, *Animal Intelligence*, Introduction.

5. On the next generation, see Edward L. Thorndike, *Animal Intelligence: Experimental Studies* (1911; reprinted New York: Hafner, 1965); Thorndike, "Do Animals Reason?" *Popular Science Monthly*, 55 (August 1899), 480–490; C. Lloyd Morgan, *Animal Behavior* (London: Edward Arnold, 1900); idem, "Limits of Animal Intelligence" (1893), reprinted in *Significant Contributions to the History of Psychology, 1750–1920*, Series D, Volume Two (Washington: University Publications of America, 1977).

6. Lindsay, *Mind in the Lower Animals*, Volume I, 52, 70, 75. The lower races of man, in contrast, were sadly lacking. In Lindsay's scheme, as one modern critic noted, the cultured English gentleman was closer to his bulldog than to the Eskimoes. See Peter H. Klopfer, *An Introduction to Animal Behavior: Ethology's First Century* (Englewood Cliffs, N. J.: Prentice-Hall, 1974), 20.

7. Romanes, *Animal Intelligence*, and *Mental Evolution in Man* (New York: D. Appleton, 1889), chart in front of book.

8. The popular literature and magazines are full of this material, and examples may be drawn from the index of almost any outdoor magazine or the shelves of older libraries. Two examples from *Outlook, 48* (September and November 1906), are "Has the Wasp Affection?" and "Do Dogs Reason?"

9. Ernest Thompson Seton, "The King of Currumpaw: A Wolf Story," *Scribner's, 16* (November 1894), 618–628; idem, *Lives of the Hunted* (New York: Charles Scribner's Sons, 1901), 12–13; idem, *Wild Animals I Have Known* (New York: Charles Scribner's Sons, 1898). The Reverend William J. Long, the main target of the "nature-faking" accusations, was the most prominent of the immediate imitators, but the popular magazines of the period are full of others now, deservedly, forgotten.

10. Ernest Thompson Seton, *Trail of an Artist-Naturalist* (New York: Charles Scribner's Sons, 1940), 6.

11. John Henry Wadland, *Ernest Thompson Seton: Man in Nature and the Progressive Era, 1880–1915* (New York: Arno, 1978); W. J. Keith, *Charles G. D. Roberts* (Toronto: Copp Clark, 1969); H. Allen Anderson, *The Chief: Ernest Thompson Seton and the Changing West* (College Station: Texas A & M University Press, 1986).

12. Seton, *Lives of the Hunted*, 12–13.

13. Seton, *Lives of the Hunted*, 274–275. For a discussion of the relation of Seton's ideas to modern ideas of animal behavior, see Wadland, *Ernest Thompson Seton*, 217–

218. Seton also wrote mammalogical studies of North American animals, in part to establish scientific credentials that would buttress the credibility of his nature fiction; see his *Life-Histories of Northern Animals* (New York: Charles Scribner's Sons, 1909), and *Lives of Game Animals* (Garden City, N.Y.: Doubleday, Page, 1925–1928).

14. Charles G. D. Roberts, *The Kindred of the Wild* (Boston: Page, 1902), 24, 28.

15. Roberts, *Kindred,* 29.

16. Roberts, *Kindred,* 29.

17. Seton, *Lives of the Hunted,* 228, 231, 251.

18. Seton, *Wild Animals I Have Known,* 11–12, 357.

19. Charles G. D. Roberts, *Watchers of the Trails* (Boston: Page, 1904), 197–208.

20. Ernest Thompson Seton, *The Ten Commandments in the Animal World* (Garden City, N.Y.: Doubleday, 1925).

21. This theme also appears in Seton's *Lives of Game Animals.*

22. Joseph Gold, "The Precious Speck of Life," *Canadian Literature,* no. 26 (Autumn 1965), 22–32. See also Gold's introduction to a collection of Roberts's work, *King of Beasts* (Toronto: Ryerson, 1967). James Polk, "Lives of the Hunted," *Canadian Literature, no. 53* (Summer 1972), 51–59. This is elaborated in Thomas R. Dunlap, " 'The Old Kinship of Earth': Science, Man, and Nature in the Animal Stories of Charles G. D. Roberts," *Journal of Canadian Studies,* 22 (Spring 1987), 104–120.

23. The story was first published in *Harper's Magazine, 86* (December 1892), 120–122. It is available in a collection, *The Last Barrier and Other Stories* (Toronto: McClelland and Stewart, 1965), 1–6.

24. The title comes from Psalm 104, verse 21, part of a psalm extolling the order of the world established by the Creator. Roberts used a quotation from a similar passage in Job 38:41 or Psalms 147:9 for the title of another early story, "The Young Ravens that Call on Him," published in *Earth's Enigmas* (1895; reprinted Boston: Page, 1902). That, too, extols the order of the world and the oversight of the Creator. W. J. Keith, "A Choice of Worlds: God, Man and Nature in Charles G. D. Roberts," in George Woodcock (editor), *Colony and Confederation: Early Canadian Poets and Their Background* (Vancouver: University of British Columbia Press, 1974), 98, discusses the titles as ironic comments, for the meat the panthers seek is man and the young ravens (eagles in the story) feed on a lamb whose cries have gone, presumably, unheard.

25. "Do Seek Their Meat from God," 122. Further discussion of Roberts's point of view is in Gold, "The Precious Speck of Life"; Gold's introduction to Roberts's work, *King of Beasts*; Polk, "Lives of the Hunted," 51–59; Robert H. MacDonald, "The Revolt Against Instinct," *Canadian Literature, no. 84* (Spring 1980), 18–29; Joseph Gold, "The Ambivalent Beast," in Carrie MacMillan, *The Proceedings of the Sir Charles G. D. Roberts Symposium* (Sackville, New Brunswick: Centre for Canadian Studies, Mount Allison University, 1984), 77–86; Keith, "A Choice of Worlds," 87–102.

26. Seton, *Wild Animals I Have Known,* 11 (italics in original).

27. Roberts, *Kindred,* 176.

28. Seton, *Lives of the Hunted,* 84.

29. Roberts, *Kindred,* 193.

30. Roberts, *Kindred,* 180–196.

31. Seton, *Wild Animals I Have Known,* 11, 267–269; idem, *Biography of a Grizzly* (New York: Century, 1900); idem, *Lives of the Hunted.*

32. Roberts, *Watchers of the Trails,* 323–348, and *The Feet of the Furtive* (New York: Macmillan, 1913), 71–94.

33. Wadland, *Ernest Thompson Seton*, 102–155.

34. Wadland, *Ernest Thompson Seton*, 122–139.

35. "Lobo" appeared in *Scribner's, 16* (November 1894), 618–628, the author signing himself Ernest E. Thompson. It appeared in book form in *Wild Animals I Have Known*, 13–44. It was reprinted as late as 1942 in *Reader's Digest, 41* (November 1942), 103–106. *Wild Animals I Have Known* is still in print. Wolves even appear as his enemies in "Bingo, the Story of a Dog." Here Seton, caught in his own wolf traps, is about to be "devoured by the foe I most despised," but he is saved by his wolf-killing dog; see *Wild Animals I Have Known*, 180.

36. Theodore Roosevelt, "Men Who Misinterpret Nature," from "Roosevelt on the Nature Fakirs," *Everybody's Magazine, 16* (June 1907), 770–774. Reprinted in *The Works of Theodore Roosevelt*, National Edition, Volume V (New York: Charles Scribner's Sons, 1926), 367–374.

37. Critical work on these stories has stressed their contribution to Canadian literature. See, for example, Margaret Atwood, *Survival: A Thematic Guide to Canadian Literature* (Toronto: Anansi, 1972), and Wadland, *Ernest Thompson Seton*, 167–175. See also Polk, "Lives of the Hunted," 51–59.

38. John Burroughs, "Real and Sham Natural History," *Atlantic Monthly, 91* (March 1903), 298–309. There is a basic bibliography on nature-faking in Loren Owings, *Environmental Values, 1860–1972* (Detroit, Mich.: Gale Research Co., 1976), 212–215. Ralph H. Lutts offered an interpretation of the nature fakers, congruent with what follows, in "The Nature Fakers: Conflicting Perspectives of Nature," in *Ecological Consciousness: Essays from the Earthday X Colloquium* (Washington: University Press of America, 1981), 183–208. See also Wadland, *Ernest Thompson Seton*, 180–264; Anderson, *The Chief*, 118–128. On the larger context of American reinterpretation of "nature" and Americans' relationship to the land, see Peter J. Schmitt, *Back to Nature: The Arcadian Myth in Urban America* (New York: Oxford University Press, 1969).

39. For examples of Long's work, see "A School for Little Fishermen," *Outing, 40* (May 1902), 150–162, and "The Partridges' Roll Call," *Outing, 40* (August 1902), 544–549. President Reagan is reported to have said that Long was his favorite childhood author; see *Washington Post*, 22 May 1987.

40. Burroughs, "Real and Sham," 301.

41. Seton, *Trail of an Artist-Naturalist*, 367–371. This version should be treated with caution.

42. Roosevelt to George Bird Grinnell, 24 April 1903, in Elting Morison, *Letters of Theodore Roosevelt*, Volume III (Cambridge: Harvard University Press, 1951), 467–470; "Hermit," "The Intelligence of the Wild Things," *Forest and Stream, 60* (18 April 1903), 304–305; William J. Long, "The Modern School of Nature-Study and Its Critics," *North American Review, 176* (May 1903), 688–698. Long received support from a Mr. Raymond S. Spears, who attacked Burroughs's own accuracy in *Forest and Stream, 60* (2 May 1903), 245–246.

43. William J. Long, "Animal Surgery," *Outlook, 75* (19 September 1903), 122–127.

44. William Morton Wheeler, "Woodcock Surgery," *Science, 19* (26 February 1904), 347–350. Replies, all in *Science, 19*, are Frank M. Chapman, "The Case of William J. Long" (4 March 1904), 387–389; W. F. Ganong, "The Writings of William J. Long" (15 April 1904), 623–625; Ellen Hayes (15 April 1904), 625–626; William J. Long, "Science, Nature, and Criticism" (13 May 1904), 760–767. The

follow-up was William Harper Davis, "Natural and Unnatural History" (22 August 1904), 667–675.

45. John Burroughs, *Ways of Nature* (Boston: Houghton Mifflin, 1905), vi, 64, 161, 171, and "Do Animals Think?" *Harper's, 110* (February 1905), 354–358.

46. Peter Rabbit (William J. Long), "Do Animals Think?" *Harper's, 111* (June 1905), 59–62, and "The Question of Animal Reason," *Harper's, 111* (September 1905), 588–594.

47. Peter Rabbit (William J. Long), "Animal Immortality," *Harper's, 111* (December 1905), 873–878.

48. Roosevelt to Burroughs, 11 June 1903, in Elting Morison, *Letters of Theodore Roosevelt*, Volume III, 486; Edward B. Clark, "Roosevelt on the Nature Fakirs," *Everybody's Magazine, 16* (June 1907), 770–774. *Outlook, 86* (8 June 1907), 263–264, chastised Roosevelt in an editorial, suggesting that he not add to the burdens of office by multiplying the problems he already faced in public life. Not daunted, Teddy fired off another rhetorical blast against the "grave wrong" committed by the "deliberate or reckless untruth" perpetuated by the fakers; see Theodore Roosevelt, "Nature Fakers," *Everybody's Magazine, 17* (September 1907), 427–430. Both reprinted in Volume V of *The Works of Theodore Roosevelt*, 367–383. Some scientists commented for the interviewer; see Edward B. Clark, "Real Naturalists on Nature Faking," *Everybody's Magazine, 17* (September 1907), 423–427.

49. *New York Times*, 2 June 1907, Sec. 5, p. 2.

50. Lorenzo P. Gibson, "A Fact or a Fake Submitted to Mr. John Burroughs," *Outing, 50* (May 1907), 553–554.

51. *New York Times*, 2 June 1907, Sec. 5, p. 1.

52. Thornton W. Burgess, *Tommy's Change of Heart* (Boston: Little, Brown, 1915), 30. Like Seton and Roberts, Burgess made a distinction between the animals, who killed to live, and humans, who hunted for sport.

53. Thornton W. Burgess, *The Burgess Animal Book for Children* (Boston: Little, Brown, 1931), 251–252.

54. A. K. Fisher, *Hawks and Owls of the United States*, Bulletin 41, Bureau of Biological Survey (Washington: Government Printing Office, 1893).

Chapter Three. Ideas and Organizations

1. Keir Sterling, *The Last of the Naturalists: The Career of C. Hart Merriam* (New York: Arno, 1977). The following account of Merriam and the Survey is drawn, from this account, unless otherwise noted.

2. Sterling, *Last of the Naturalists*, 63–70; A. Hunter Dupree, *Science in the Federal Government* (Cambridge: Harvard University Press, 1957), Chapter 8.

3. Teddy Roosevelt was to recall that when he went to college his interest in natural history was quickly choked off by the exclusive emphasis on laboratory studies and specialized scholarship; see *The Autobiography of Theodore Roosevelt* (New York: Charles Scribner's Sons, 1926), 25–27. See also Ronald Tobey, *Saving the Prairies* (Berkeley: University of California Press, 1981), Chapters 1 and 2.

4. *American Men of Science* found in 1903 that about three-fourths of the leading scientists had bachelor's degrees and half had the Ph.D; see "A Further Statistical Study of American Men of Science," *American Men of Science*, 1910 edition, 564–596. Merriam's cohort comes from lists in Survey publications, from the reminiscences of Tracy Storer, "Mammalogy and the American Society of Mammalogists," *Journal of Mammalogy, 50* (November 1969), 785–793, and from Sterling, *Last of the*

Naturalists. Merriam's own opinions are in C. Hart Merriam, "Biology in Our Colleges: A Plea for a Broader and More Liberal Biology," *Science, 21* (30 June 1893), 353–355. Versions of this appeared in *Public Opinion, 15* (15 July 1893), 344–345, and the *Utica* [New York] *Herald,* 15 August 1893. See also reply to Merriam in *Science, 22* (14 July 1893), 21–22.

5. By the early twentieth century its studies of the diet of birds and mammals (it had data from more than 50,000 bird stomachs) made the Survey the most important source of information on this subject in the world. Harold C. Bryant, "A Determination of the Economic Status of the Western Meadow-Lark (Sturnella neglecta) in California," *University of California Publications in Zoology, 11* (1914), 388, provides a brief introduction. Most of this work was done under the direction of W. L. McAtee, who joined the Survey in 1904. See McAtee Papers, Library of Congress, Manuscript Division.

6. On agency building in the federal government, the issue here, see Dupree, *Science in the Federal Government,* Chapter 8. On the Biological Survey, see Sterling, *Last of the Naturalists.* The annual *Report of the Secretary of Agriculture* (Washington: Government Printing Office, annual) provides valuable data. Agency links to academia are apparent not only there but in the Correspondence files of the Museum of Vertebrate Zoology, Berkeley, California.

7. On organization, see Sterling, *Last of the Naturalists,* 308. Several Survey heads, employees, or ex-employees, became president, and a Survey worker, Viola Schantz, served as secretary for years.

8. McAtee Papers, Library of Congress, Box 80, Memos of 1935, 1937, 1938; W. L. McAtee, "Long Live the Wildlife Society," *Journal of Wildlife Management, 1* (July 1937), 45–46; idem, "Notes," *Journal of Wildlife Management, 2* (April 1938), 61; Victor H. Cahalane, "Report of the Secretary," *Journal of Wildlife Management, 2* (April 1938), 65–68.

9. On the development of administrative agencies based on science, see Samuel P. Hays, *Conservation and the Gospel of Efficiency* (Cambridge: Harvard University Press, 1959). On the role of hunting, see John F. Reiger, *American Sportsmen and the Origins of Conservation* (New York: Winchester, 1975). On wildlife law, see Michael Bean, *The Evolution of National Wildlife Law* (New York: Praeger, 1983); James Tober, *Who Owns the Wildlife? The Political Economy of Conservation in Nineteenth-Century America* (Westport, Conn.: Greenwood, 1981); Thomas Lund, *American Wildlife Law* (Berkeley: University of California Press, 1980); T. S. Palmer, "Chronology and Index of the More Important Events in American Game Protection, 1776–1911," Bulletin 41, Bureau of Biological Survey (Washington: Government Printing Office, 1912).

10. Bean, *National Wildlife Law,* Chapter 2. This seems far-fetched, and the 1916 treaty with Canada was an attempt to place this authority on firmer ground, but the commerce clause was upheld as a legitimate basis for regulation in several cases. See Bean, *National Wildlife Law,* 78–79; Cases 6174, 6175 appealed from U.S. District Court for the southern District of Ilinois, Southern Division, to 7th Circuit Court of Appeals, affirmed 4 October 1937. Documents and Canadian discussion in Acts and Legislation, Box 3, Record Group 109, Records of the Canadian Wildlife Service, Public Archives Canada, Ottawa (these records hereafter cited as RG 109, Public Archives Canada).

11. Bean, *National Wildlife Law,* 19–21. On the treaty, see Janet Foster, *Working for Wildlife: The Beginning of Preservation in Canada* (Toronto: University of Toronto Press, 1978), 132–145. Canadian archives show that much the same argument was

used in Canada. Game was, under the British North America Act, the responsibility of the province, but federal treaty powers overrode provincial law. See Acts and Legislation, Boxes 3 and 4, RG 109, Public Archives Canada.

12. Palmer, "Chronology and Index"; idem, "Extermination of Noxious Animals by Bounties," in U.S. Department of Agriculture, *Yearbook of the United States Department of Agriculture, 1896* (Washington: Government Printing Office, 1897), 55–68; David E. Lantz, "Use of Poisons for Destroying Noxious Animals," in U.S. Department of Agriculture, *Yearbook of the United States Department of Agriculture, 1908* (Washington: Government Printing Office, 1908), 421–432; idem, "Destroying Prairie-Dogs and Pocket Gophers," Bulletin 116, Experiment Station, Kansas State Agricultural College (Manhattan, Kansas: Kansas State Agricultural College, 1903); "Predatory—Sale of Poisons," General Correspondence, 1890–1944, Bureau of Biological Survey, Fish and Wildlife Service, Record Group 22, Records of the United States Fish and Wildlife Service, National Archives, Washington, D.C. (hereafter cited as RG 22, National Archives), contains correspondence on state poison policies.

13. Vernon Bailey, "Wolves in Relation to Stock, Game, and the National Forest Reserves," Forest Service Bulletin 72 (Washington: Government Printing Office, 1907).

14. "Report on Work of the Biological Survey," Senate Document 132, 60th Congress, 1st Session. Copy from C. Hart Merriam Papers, Manuscript Division, Library of Congress, Box 53. The activities of the Bureau may be best traced through Bureau of Biological Survey, *Annual Report of the Chief of the Bureau of Biological Survey* (Washington: Government Printing Office, annual).

15. Files in the General Records of the Fish and Wildlife Service, RG 22, National Archives, contain unpublished recollections of the early years. Stanley Paul Young (undated paper on the history of predator and rodent control work, Folder 11, Box 7, Young Papers, Smithsonian Institution Archives, Washington, D.C.), described experiments in poisoning prairie dogs in South Park, Colorado, in 1912. Annual reports of the Survey indicate growing awareness of the problem and Merriam's attempts to meet it, as do the office memos from this period in the McAtee Papers, Manuscript Division, Library of Congress.

16. Jenks Cameron, *The Bureau of Biological Survey* (Baltimore: Johns Hopkins Press, 1929; reprinted New York: Arno, 1974), 45–46.

17. The Division went through several name changes. This account uses the common abbreviation PARC, even when the name was something different. Name changes in the last twenty years are indicated in the text as they occur. A full list is in Charles L. Cadieux, *Coyotes: Predators and Survivors* (Washington: Stone Wall Press, 1983), 231–232.

18. Stanley Paul Young, *The Last of the Loners* (New York: Macmillan, 1970).

19. Office Memos, 1925 and later, document this change. McAtee Papers, Manuscript Division, Library of Congress.

20. Vernon Bailey Papers, Folder 12, Box 5, Smithsonian Institution Archives, Washington, D.C.

21. Bailey, "Wolves in Relation to Stock and Game"; Bailey Papers, Smithsonian; and "Balance of Nature," General Files, Division of Wildlife Services, RG 22, National Archives.

22. Young, *Last of the Loners*, iv.

23. Young Papers, Conservation Library, Denver Public Library, Denver, Colorado.

24. Stanley Paul Young, "The War on the Wolf," Part II, *American Forests, 48* (December 1942), 574.

25. Dupree, *Science in the Federal Government*, Chapter 8, discusses the development of this network of institutions; Tobey, *Saving the Prairies*, describes the close links between plant ecology and the applied work of the experiment stations. Leland O. Howard, one of the more successful agency builders in the USDA, described the effect of the Hatch Act on applied biologists in *A History of Applied Entomology*, Smithsonian Institution Miscellaneous Publications, Volume 84 (Washington: Government Printing Office, 1931), 74–75, 105–114. See also Eugene Cittadino, "Ecology and the Professionalization of Botany in America, 1890–1905," *Studies in the History of Biology, 4* (1980), 174–175.

26. Frank Egerton (editor), *History of American Ecology* (New York: Arno, 1977); idem, "The History of Ecology: Achievements and Opportunities, Part One," *Journal of the History of Biology, 16* (Summer 1983), 259–310; idem, "The History of Ecology: Achievements and Opportunities, Part Two," *Journal of the History of Biology, 18* (Spring 1985), 103–143.

27. Stephen Forbes, "The Lake as Microcosm," *Illinois Laboratory of Natural History Bulletin, 15* (1887), reprinted in Keir Sterling (editor), *Ecological Investigations of Stephen Alfred Forbes* (New York: Arno, 1977); Howard, *A History of Applied Entomology.*

28. Stephen A. Forbes, "The Regulative Action of Birds Upon Insect Oscillations," *Illinois Laboratory of Natural History Bulletin, 1* (1883), reprinted in Sterling (editor), *Ecological Investigations of Forbes.*

29. Leland O. Howard, "Danger of Importing Insect Pests," in U.S. Department of Agriculture, *Yearbook of the United States Department of Agriculture, 1897* (Washington: Government Printing Office, 1898), 529–552; Charles Elton, *The Ecology of Invasions by Plants and Animals* (London: Methuen & Co., Ltd., 1958); Leland O. Howard and W. F. Fiske, "The Importation into the United States of the Parasites of the Gypsy Moth and the Brown-Tail Moth," Bulletin 91, USDA Bureau of Entomology (Washington: Government Printing Office, 1911); W. D. Hunter and W. D. Pierce, "The Mexican Cotton Boll Weevil: A Summary of the Investigations of This Insect up to December 31, 1911," Bulletin 114, USDA Bureau of Entomology (Washington: Government Printing Office, 1912). On the importance of these for ecology, see Egerton, "History of Ecology, Parts One and Two" and the introductory historical chapter in W. C. Allee, Orlando Park, Alfred E. Emerson, Thomas Park, and Karl P. Schmidt, *Principles of Animal Ecology* (Philadelphia: W. B. Saunders, 1949).

30. John C. Phillips, "Wild Birds Introduced or Transplanted in North America," U.S. Department of Agriculture, Bulletin 61 (Washington: Government Printing Office, 1928); Gardiner Bump, "The Introduction and Transplantation of Game Birds and Mammals into the State of New York," in *Transactions of the Fifth North American Wildlife Conference* (Washington: American Wildlife Institute, 1941), 409–420; Robin Doughty, *Wildlife and Man in Texas* (College Station: Texas A & M University Press, 1983), Chapter 5; Elton, *Ecology of Invasions.*

31. T. S. Palmer, "The Danger of Introducing Noxious Animals and Birds," in U.S. Department of Agriculture, *Yearbook of the United States Department of Agriculture, 1898* (Washington: Government Printing Office, 1899), 93.

32. Howard and Fiske, "Importation"; Hunter and Pierce, "Mexican Cotton Boll Weevil"; Thomas R. Dunlap, "The Gypsy Moth: A Study in Science and Public Policy," *Journal of Forest History, 24* (July 1980), 116–126.

33. Frank N. Egerton, "Changing Concepts of the Balance of Nature," *Quarterly Review of Biology, 48* (June 1973), 320–350; Donald Worster, *Nature's Economy* (San

Francisco: Sierra Club Books, 1977; reprinted New York: Cambridge University Press, 1985).

34. Howard and Fiske, "Importation"; Hunter and Pierce, "Mexican Cotton Boll Weevil"; Allee et al., *Principles of Animal Ecology*, Chapter 1.

35. Charles Elton, *Animal Ecology* (London: Sidgwick and Jackson, 1927), 3.

36. Other scientists criticized the scheme for pinning too much on an imprecise variable—temperature. There were problems even in deciding what temperature to use (highest, highest average, lowest, lowest average) and, critics complained, too many exceptions to the boundaries.

37. For a critique of this theory and of its predecessors, see Keir Sterling's *The Last of the Naturalists*, Chapter 6.

38. On Clements and the prairie ecologists, see Tobey, *Saving the Prairies*. He argues for an intimate connection of the Clementsian tradition to the land-grant conditions of the university. I find this case less than compelling.

39. Tobey, *Saving the Prairies*. The climax theory was given definitive form in Clements's *Plant Succession* (Washington: Carnegie Institution, 1916).

40. Charles C. Adams, *A Guide to the Study of Animal Ecology* (New York: Macmillan, 1913).

41. Victor Shelford, *Animal Communities of North America* (Chicago: University of Chicago Press, 1917).

42. On early work, see Cittadino, "Ecology and Professionalization"; on the Ecological Society of America, see Robert L. Burgess, "The Ecological Society of America: Historical Data and Some Preliminary Analyses," in Egerton (editor), *History of American Ecology*.

43. On the development of these areas, see Robert P. McIntosh, *The Background of Ecology: Concept and Theory* (New York: Cambridge University Press, 1985); on mathematical ecology, see Sharon Kingsland, *Modeling Nature* (Chicago: University of Chicago Press, 1985).

44. Shelford, *Animal Communities*, 1.

45. Cittadino, "Ecology and Professionalization," 182; Adams, *Animal Ecology*, 3–5.

Chapter Four. Worthless Wildlife

1. Speech opening a conference of field agents in Ogden, Utah, 23 April 1928. "Report of Conferences, 1928–1941," General Files, Division of Wildlife Services, Records of the United States Fish and Wildlife Service, Record Group 22, National Archives, Washington, D.C. (hereafter cited as RG 22, National Archives).

2. Chapter 13 of Donald Worster's *Nature's Economy* (San Francisco: Sierra Club Books, 1977; reprinted New York: Cambridge University Press, 1985) is entitled "The Value of a Varmint." I have borrowed more than the chapter title; much of this chapter is devoted to proving that Worster's view of this incident is not entirely correct.

3. See *Annual Report of the Secretary of Agriculture* (Washington: Government Printing Office, annual).

4. E. A. Goldman, "The Coyote—Archpredator," *Journal of Mammalogy*, 11 (August 1930), 325–334.

5. Counts are in annual reports. C. C. Presnall sent a yearly summary to J. Frank Dobie, 24 January 1947; in Coyote correspondence, RG 22, National Archives. The counts are not accurate, for many of the same reasons that body counts in Vietnam

bore little relation to reality. Coyotes might also die where the hunter would not find them. The figures, however, indicate considerable effort and tangible results.

6. Joseph Grinnell to C. Hart Merriam, 27 November 1907, Box 78, W. L. Mc-Atee Papers, Manuscript Division, Library of Congress. Joseph Grinnell, "The Methods and Uses of a Research Museum," *Popular Science Monthly,* 77 (August 1910), 31–40, reprinted in a posthumous collection, *Joseph Grinnell's Philosophy of Nature* (Freeport, N. Y.: Books for Libraries, 1943). On the vanishing West of this period, see Lee Clark Mitchell, *Witnesses to a Vanishing America: The Nineteenth-Century Response* (Princeton: Princeton University Press, 1981). On western leadership, see H. E. Anthony to Joseph Grinnell, 3 June 1925, Anthony file, Correspondence of the Museum of Vertebrate Zoology, Berkeley, California, cited with permission of David Wake, Director (hereafter cited as MVZ). Evidence of the mammalogists' feelings comes principally from correspondence files of Joseph Grinnell, E. Raymond Hall, Joseph Dixon, C. C. Adams, Lee R. Dice, H. E. Anthony, A. Brazier Howell, and Ira N. Gabrielson, MVZ. See also "Predatory—Sale of Poisons" file in General Correspondence, 1890–1944, and "Predatory—States" file in General Files, Division of Wildlife Services, RG 22, National Archives.

7. Hall file, other workers' files, Correspondence, MVZ.

8. Correspondence, MVZ; *Joseph Grinnell's Philosophy of Nature.*

9. Ira N. Gabrielson to ASM President Witmer Stone, 24 March 1931, and later correspondence, Anthony file, MVZ. See also A. Brazier Howell file, Department of Mammalogy, American Museum of Natural History, New York.

10. Participants included H. E. Anthony, "General Status of Predatory Mammal Problems"; Lee R. Dice, "Scientific Value of Predatory Mammals"; W. B. Bell, "Predatory Mammals, a Practical Problem in Economics"; E. A. Goldman, "The Predatory Mammal Problem and the Balance of Nature"; Joseph Dixon, "Food Predilections of Predatory Mammals"; and C. C. Adams, "The Conservation of Predatory Mammals." The papers of Dice, Goldman, Dixon, and Adams were printed in the *Journal of Mammalogy,* 6 (February 1925). Bell's is in Research Reports, General Files, Division of Wildlife Research, RG 22, National Archives.

11. Adams, "Conservation"; Dixon, "Food Predilections"; Dice, "Scientific Value."

12. Goldman, "The Predatory Mammal Problem."

13. Jenks Cameron, *The Bureau of Biological Survey* (Baltimore: Johns Hopkins University Press, 1929; reprinted New York: Arno, 1974), 45–46. The Survey's scientific work, it seemed, was also going to the dogs. See W. L. McAtee to Grinnell, 7 July 1931, McAtee file, MVZ, and A. B. Howell to Grinnell, 1926, Howell file, MVZ.

14. Joseph Dixon and E. W. Nelson files, MVZ.

15. David E. Lantz, "Coyotes in Their Economic Relation, Bureau of Biological Survey, Bulletin 20 (Washington: Government Printing Office, 1905), 16–17. Vernon Bailey, "Wolves in Relation to Stock, Game, and the National Forest Reserves," Forest Service Bulletin 72 (Washington: Government Printing Office, 1907); W. B. Bell, "Hunting Down Stock Killers," in U.S. Department of Agriculture, *Yearbook of Agriculture, 1920* (Washington: Government Printing Office, 1921), 295–296; idem, "Predatory Animals—A Problem in Economics," Bell file, Research Reports, Division of Wildlife Research, Records of the Fish and Wildlife Service, RG 22, National Archives, 17. E. Raymond Hall wrote Bell (7 October 1921, Bell file, MVZ) asking about evidence for damage estimates. There is no reply, and given the meticulous records of the MVZ, it is likely there was none. On estimates, see also Merriam Papers, Box 53, Manuscript Division, Library of Congress.

16. T. S. Palmer, "Extermination of Noxious Animals by Bounties," in U.S. De-

partment of Agriculture, *Yearbook of the United States Department of Agriculture, 1896* (Washington: Government Printing Office, 1897), 55–68, gives information on bounties. Ora B. Peake, *The Colorado Range Cattle Industry* (Glendale, Calif: Arthur H. Clark, 1937), 231–243, gives figures for Colorado. On large bounties for individual outlaw wolves, see Stanley Paul Young, *The Last of the Loners* (New York: Macmillan, 1970).

17. "Report of Conferences, 1928–1941," RG 22, National Archives.

18. E. Raymond Hall, "An Outbreak of House Mice in Kern County, California," *University of California Publications in Zoology, 30* (1927), 189–203; Hall field notes, MVZ.

19. Hall field notes, MVZ; Stanley E. Piper, "The Mouse Infestation of Buena Vista Lake Basin," in Department of Agriculture, State of California, *Monthly Bulletin, 17* (October 1928), 538–560. On the "balance of nature" argument, see Frank N. Egerton, "Changing Concepts of the Balance of Nature," *Quarterly Review of Biology, 48* (June 1973), 322–350; "Balance of Nature" file, General Files, Division of Wildlife Services, RG 22, National Archives.

20. The thrust of his views (though he would have indignantly rejected the proposition, stated so baldly) was that predation regulated populations, that the physical environment had no significant effect. Hall, still active when I interviewed him in the summer of 1982, believed that he had, in the 1920s, sufficient scientific information to show that the PARC's program was destructive to normal wildlife populations. A good but impressionistic study comes a decade later in Laurence M. Huey, "El Valle De La Trinidad, the Coyote Poisoner's Proving Ground," *Journal of Mammalogy, 18* (February 1937), 74–76.

21. Even professionally, ecology was in a state of flux. Some of the protestors belonged to the Ecological Society of America, but it was the American Society of Mammalogists that took the lead in fighting the predator control program; see Robert L. Burgess, "History of the Ecological Society of America, 1977," in Frank Egerton (editor), *History of American Ecology* (New York: Arno, 1977), gives a history of the early organization and activities of the Society. The Barrington Moore file, MVZ, contains more on the ecologists' views.

22. Joseph Grinnell, "Significance of Faunal Analysis for General Biology," *University of California Publications in Zoology, 32* (November 1928), 13–18.

23. Tracy Storer, *From Observation to Experiment*, oral interview, University of California Archives, Davis, California, 42, said that Grinnell balked at one of his projects, which, had he been able to do it, would have given him "an early entry into the field of ecology." Historian Alfred Runte, personal communication, suggested that Grinnell's enthusiasm for his students' ecological work varied with the degree of credit they gave him for their inspirations. See Barrington Moore file, MVZ, on Grinnell's interest in ecology. On the niche, see John H. Vandermeer, "Niche Theory," *Annual Review of Ecology and Systematics, 3* (1972), 107; Robert H. Whittaker and Simon A. Levin (editors), *Niche: Theory and Applications* (Stroudsburg, Pa.: Dowden, Hutchinson, and Ross, 1975), 2–11; and Robert P. McIntosh, *The Background of Ecology: Concept and Theory* (New York: Cambridge University Press, 1985), 190–191, 281–282.

24. Biographical material from *American Men of Science* and Henry Clepper, *Leaders in American Conservation* (New York: Ronald Press, 1971).

25. "Report of the Committee on Wild Life Sanctuaries, Including Provision for Predatory Mammals," *Journal of Mammalogy, 9* (November 1928), 354–358.

26. Paul G. Redington, "Policy of the U.S. Biological Survey in Regard to Preda-

tory Mammal Control," *Journal of Mammalogy, 10* (August 1929), 276–279. Adams's letter precedes this article. Research, despite Redington's claims, was clearly less important than control; there were complaints at field agent conferences that the men could not be expected to do scientific work in addition to running their trap lines and bait stations; see "Report of Conferences, 1928–1941," General Files, Division of Wildlife Services, RG 22, National Archives.

27. Paul G. Redington, speech opening conference of field agents in Ogden, Utah, 23 April 1928; "Report of Conferences, 1928–1941," General Files, Division of Wildlife Services, RG 22, National Archives.

28. The files of the Bureau of Biological Survey, RG 22, are filled with such references. U.S. House of Representatives, *Agricultural Department Appropriations Bill for 1933*, Hearing before the Subcommittee of House Committee on Appropriations, 72nd Congress, 2nd Session (Washington: Government Printing Office, 1932).

29. Cameron, *Biological Survey*; C. C. Adams, "Rational Predatory Animal Control," *Journal of Mammalogy, 11* (August 1930), 353–358. Goldman's comment is on p. 360 in the discussion following Adams's paper. The editor's intemperate outburst against the distortions of reporters is in *Journal of Mammalogy, 19* (November 1929), 373.

30. Copies of the petition and correspondence are in Howell file, MVZ. See also "Predatory—States, Maryland," General Files, Division of Wildlife Services, RG 22, National Archives.

31. On the matter of food habits and the trapping of nontarget species, we have the testimony of Tracy I. Storer, who visited Washington in the spring of 1930 while Congress was holding hearings on the predator control program. His comments on the Survey are the more valuable in that he was not only a neutral observer (so far as that could be said of anyone who was involved, even peripherally) but he confided his comments to his private field notes (which he kept from 1912 to 1963); they were never intended for publication. On 2 May 1930 he had lunch with W. C. McAtee and visited the food habits laboratory. "When I enquired about mammal stomachs I was told that very few had been examined. I was shown data sheets or rather reports received from the 'leaders' in predator animal control in several western states. The food examinations are made chiefly by trappers. From 25 to 33% is 'unidentified.' There were very few records made of species other than coyotes." Storer then gave figures indicating almost no collections. "Obviously skunks are not being reported since it is practically impossible in the Western states to get coyotes without numerous skunks, by either trapping or poisoning. Dr. A. K. Fisher came in while this discussion was in progress and endeavored to defend present practices but he could not cite any work done on food habits of mammals in the predatory class." Unpublished field notes of Tracy I. Storer, 1912–1963, 1232, used with permission of Dr. Ruth Risdon Storer, 619 Oak Avenue, Davis, California. W. C. Henderson, "The Control of the Coyote," *Journal of Mammalogy, 11* (August 1930), 336–350, is a good example of the Survey's use of its statistics. The printed paper includes the discussion and criticism that followed its presentation to the mammalogists.

32. This was a common complaint from the mammalogists, who distrusted the Survey's hunters. See Howell letter of 14 April 1930, Howell file, MVZ, Berkeley, and other files in this collection, particularly H. E. Anthony and E. Raymond Hall; see also Storer comments in note 31.

33. A. Brazier Howell, "The Poison Brigade of the Biological Survey," *Outdoor*

Life, 66 (July 1930), 32–33; idem, "The Borgias of 1930," *Outdoor Life, 66* (September 1930), 17–20; W. C. Henderson, "The Other Side of the Poison Case," *Outdoor Life, 66* (December 1930), 22–23; E. Raymond Hall, "The Poisoner Again," *Outdoor Life, 67* (April 1931), 26–27; 67 (May 1931), 28–29; H. E. Anthony, "The Control of Predatory Mammals," *Science, 74* (18 September 1931), 288–290; A. Brazier Howell, "Poisoning Rodents—and Then," *Science, 74* (18 December 1931), 632; E. A. Goldman, "The Control of Injurious Animals," *Science, 75* (18 March 1932), 309–311.

34. U.S. Congress, House of Representatives, Committee on Agriculture, *Control of Predatory Animals,* House Document 496, 70th Congress, 2nd Session (Washington: Government Printing Office, 1929); idem, *Control of Predatory Animals,* Hearings before the Committee on Agriculture, 71st Congress, 2nd Session (Washington: Government Printing Office, 1930); idem, *Predatory and Other Wild Animal Control,* House Report 2396 to accompany H.R. 9599, 71st Congress, 3rd Session (Washington: Government Printing Office, 1931); U.S. Congress, Senate, Committee on Agriculture and Forestry, *Control of Predatory Animals,* Hearings before the Committee on Agriculture and Forestry on S. 3483, 71st Congress, 2nd and 3rd Sessions (Washington: Government Printing Office, 1931); Hall file, MVZ. Storer's field notes show that there was considerable friction and much "amending" of the record before it was printed. Hall's correspondence for the period indicates the same.

35. Goldman, "Archpredator"; Henderson, "Control"; E. Raymond Hall, "Predatory Mammal Destruction," *Journal of Mammalogy, 11* (August 1930), 373–377; Adams, "Rational Control"; A. Brazier Howell, "At the Cross-Roads," *Journal of Mammalogy, 11* (August 1930), 377–389. The resolution was printed in *Journal of Mammalogy, 11* (August 1930), 431.

36. Howell to Hall, 5 September 1930, Howell file, MVZ. In his reply, 17 September 1930 (same file), Hall said that trappers had told him "confidentially that they, of course, would not be able to state their views or experiences to anyone connected with the Biological Survey," or anyone who might report it there, "for it would jeopardize their position." Some of the trappers "intimated . . . that they were convinced that the work was all wrong from an economic point of view" but were afraid if they spoke out they would be blacklisted. See also Dixon, Anthony, Adams, and Dice files, MVZ; correspondence of O. J. Murie, who headed a regional office for the PARC in Jackson, Wyoming, in "Report on Poisoning" file, Predatory Animals, Murie Papers, Denver Conservation Library, Denver, Colorado; "Predatory—Sale of Poisons," General Correspondence, RG 22, National Archives. The accusation of "drumming up" business recurred; see Clarence Cottam to J. T. McBroom, 16 January 1967, in "Policy—Correspondence re Control Policy," General Records, Division of Wildlife Services, RG 22, National Archives.

37. James C. Munch and James Silver, "The Pharmacology of Thallium and Its Use in Rodent Control," USDA Technical Bulletin 238 (Washington: Government Printing Office, 1931), 1–2, 21. See also Thallium files, General Files, Division of Wildlife Services, RG 22, National Archives.

38. Joseph Grinnell to E. W. Nelson, Chief of the Survey, 2 December 1926, reporting on a letter to him from a Mr. Morgan Hill of Santa Clara County, in Nelson file, MVZ. See also "Report on Conferences, 1928–1941" file, in Records of the Bureau of Biological Survey, RG 22, National Archives, and Munch and Silver, "Use of Thallium."

39. Jean M. Linsdale, "Facts Concerning the Use of Thallium in California to

Poison Rodents—Its Destructiveness to Game Birds, Song Birds and Other Valuable Wildlife," *Condor, 33* (May 1931), 92–106; Joseph Grinnell, "Wholesale Poisoning of Wild Animal Life," *Condor, 33* (May 1931), 131–132.

40. Annie M. Alexander, "Control, Not Extermination, of *cynomys Ludovicianus Arizonensis*," *Journal of Mammalogy, 13* (August 1932), 302; T. Gilbert Pearson, "Poisoning Birds and Mammals," *Bird-Lore, 33* (September-October 1931), 362–364. On the Survey's immediate response, see G. E. Garlow to Stanley Paul Young, 16 July 1931, and other letters in "Predatory—States, New York" file, attached to correspondence with Madison Grant, in RG 22, National Archives.

41. Eugene S. Kellogg, "The California Ground Squirrel Control Program," Special Publication 109, California Department of Agriculture (Sacramento: State Printing Office, 1931). Linsdale's rebuttal is in "Further Facts Concerning Losses to Wild Animal Life Through Pest Control in California," *Condor, 34* (May 1934), 121–135. See also Young file, MVZ.

42. Johnson Neff, notes on a meeting of the Cooper Ornithological Club, 24 November 1931, Office Memos, 1931, Box 80, McAtee Papers, Manuscript Division, Library of Congress.

43. (46 Stat. 1468; 7 U.S.C. 426-426b); discussion in Michael Bean, *The Evolution of National Wildlife Law* (New York: Praeger, 1983), 235–236.

44. For protests by the mammalogists and other groups, see the files "Predatory—Sale of Poisons" and "Emergency Conservation Work" in General Correspondence, RG 22, National Archives.

45. Gabrielson file, MVZ; Anthony to Howell, 9 May 1936, Howell file, Department of Mammalogy, American Museum of Natural History, New York.

46. O. J. Murie to W. C. Henderson, 9 January 1931, Box 265, Predatory Animals, Report on Poisoning file, Murie Papers, Conservation Center, Denver Public Library, Colorado.

47. Murie to A. Brazier Howell, 7 May 1931, Department of Mammalogy, American Museum of Natural History, New York. Murie was not alone; see K. C. McMurray to J. N. Darling (head of Survey), 18 December 1934, in "Predatory—States, Michigan," General Files, RG 22, National Archives.

48. O. J. Murie to C. C. Presnall, 7 December 1952, in Miscellaneous P file, Murie Papers, Denver Public Library.

49. Vernon Bailey, "The Home Life of the Big Wolves," *Natural History, 46* (September 1940), 120–122.

Chapter Five. Making a New Wildlife Policy

1. Mention of the Kaibab is almost universal in discussions of deer problems and hunting in the popular press and in textbooks. For quantitative counts in texts, see C. John Burk, "The Kaibab Deer Incident: A Long-persisting Myth," *Bioscience, 23* (February 1973), 113–114, and Frederic H. Wagner, "Livestock Grazing and the Livestock Industry," in Howard P. Brokaw (editor), *Wildlife in America* (Washington: Government Printing Office, 1978), 121–145. A late example is James B. Trefethen, "The Terrible Lesson of the Kaibab," *National Wildlife, 5* (June-July 1967), 4–9. Graeme Caughley disputed the conventional wisdom in "Eruption of Ungulate Populations, with Emphasis on Himalayan Thar in New Zealand," *Ecology, 51* (Winter 1970), 53–72. In April 1987 the *Journal of Forest History* accepted my article on popular and scientific reaction to the Kaibab.

2. John P. Russo, *The Kaibab North Deer Herd: Its History, Problems, and Management,*

Wildlife Bulletin Number Seven, Arizona Game and Fish Department (Phoenix, Arizona: State Game and Fish Department, 1964), 125; Walter G. Mann and S. B. Locke, "The Kaibab Deer: A Brief History and Recent Developments," Forest Service, USDA (May 1931), copy from E. Raymond Hall, Dyche Museum of Natural History, University of Kansas, Lawrence, Kansas. For a more extensive record, see Entry 73 (Records of the Division of Wildlife Management), Correspondence, Region 3, Kaibab, in Records of the United States Forest Service, Record Group 95, National Archives, Washington, D.C. (hereafter cited as Kaibab, RG 95, National Archives). This includes R. E. Ratchford, "History of the Kaibab Deer Herd," compiled January 1931, and copies of most official reports, requests, and correspondence. Aldo Leopold's famous essay, "Thinking Like a Mountain," in *A Sand County Almanac* (1948; reprinted New York: Ballantine, 1969), 137–141, is based on this experience.

3. Mann and Locke, "Kaibab Deer"; Ratchford, "Kaibab Deer Herd."

4. William Rush, "Wildlife Census," *American Forests, 48* (January 1942), 20.

5. Ratchford, "Kaibab Deer Herd."

6. On wildlife law, see Michael Bean, *The Evolution of National Wildlife Law* (revised edition, New York: Praeger, 1983), Chapter 2, and James C. Foster, "The Deer of Kaibab: Federal-State Conflict in Arizona," *Arizona and the West, 12* (Autumn 1970), 255–268. Early consideration of constitutional questions comes in a memo from P. C. Knox, Attorney General of the United States, to the Honorable John F. Lacey, House of Representatives, 3 January 1902; an answer to Lacey's request of 5 December 1901, is in Box 18, Kaibab, RG 95, National Archives.

7. E. W. Nelson to George Bird Grinnell, 5 November 1923, copy marked "Personal and Confidential," Box 225, General Correspondence Relating to Wildlife Management, 1890–1956, Reservations, National Forests, Kaibab, Record Group 22, Records of the United States Fish and Wildlife Service, National Archives, Washington, D.C. (hereafter cited as Wildlife Management, RG 22, National Archives).

8. See the popular journals of the period, particularly *Forest and Stream, Recreation*, and *Outdoor Life*.

9. C. E. Ratchford to William Greeley, 24 November 1922, Box 18, Kaibab, RG 95, National Archives. The Forest Service faced similar problems of reconciling scientific management with public opinion in other areas, particularly forest fire control policies. See Ashley Schiff, *Fire and Water* (Cambridge: Harvard University Press, 1962), and Stephen J. Pyne, *Fire in America* (Princeton: Princeton University Press, 1982), 100–122.

10. It consisted of Hayward Cutting of the Boone and Crockett Club, T. W. Tomlinson of the American National Livestock Breeders Association, John C. Burnham of the American Game Protective Association, and T. Gilbert Pearson of Audubon.

11. Mann and Locke, "Kaibab Deer," 13; T. Gilbert Pearson, *Adventures in Bird Protection* (New York: D. Appleton-Century, 1937), 316–317.

12. Leopold, *Sand County Almanac*, 138, 140. For a discussion of Leopold's thought, see Susan Flader, *Thinking Like a Mountain* (Columbia, Missouri: University of Missouri Press, 1974). Flader's excellent study has been a major influence on my own ideas.

13. Foster, "The Deer of Kaibab"; Forest Service correspondence, Kaibab, RG 95, National Archives; Box 225, Wildlife Management, RG 22, National Archives.

14. 31 December 1924, *Tucson Star*, clipping in Box 16, Kaibab, RG 95, National Archives; Foster, "The Deer of Kaibab."

15. Documents in Ratchford, "Kaibab Deer Herd," give a good account of this maneuvering.

16. Ratchford, "Kaibab Deer Herd," 140–150. Case is 278 U.S. 96 (1928), quoted in Bean, *Wildlife Law*, 22. A later decision, *Chalk v. United States* (114 F.2d 207 [4th Cir. 1940]), extended the *Hunt* decision, and in the last twenty years a series of cases had further expanded federal control over resident wildlife on federal land. Bean, *National Wildlife Law*, 21–28; Foster, "Deer of Kaibab"; Forest Service correspondence, Kaibab, RG 95, National Archives.

17. Barnes to Ratchford, 1 November 1927, Box 16, Kaibab, RG 95, National Archives.

18. Phoenix (Arizona) *Gazette*, 9(?) October 1930, clipping from Box 16, Entry 73, RG 95, National Archives.

19. Aldo Leopold, Lyle K. Sowls, and David L. Spencer, "A Survey of Over-Populated Deer Ranges in the United States," *Journal of Wildlife Management*, *11* (April 1947), 162–177.

20. Pearson, *Adventures in Bird Protection*, 321–323; E. Raymond Hall, field notes, Records of the Museum of Vertebrate Zoology, 8 June to 15 June 1931, used with the permission of David Wake, Director; Entry 73, RG 95, National Archives.

21. Nelson to Grinnell, 5 November 1923, Box 225, Wildlife Management, RG 22, National Archives.

22. Entry 73, RG 95, National Archives.

23. R. P. Boone, "Deer Management on the Kaibab," in American Wildlife Institute, *Transactions of the Third North American Wildlife Conference* (Washington: American Wildlife Institute, 1938), 368–375.

24. Charles Elton, *Animal Ecology* (London: Sidgwick and Jackson, 1927; reprinted London: Meuthen, 1966). Quote from 1966 edition, vii.

25. On the interaction between applied and "pure" studies in this area, see the introductory historical chapter in W. C. Allee, Orlando Park, Alfred E. Emerson, Thomas Park, and Karl P. Schmidt, *Principles of Animal Ecology* (Philadelphia: W. B. Saunders, 1949), and David Lack, *The Natural Regulation of Animal Numbers* (London: Oxford University Press, 1954), 1, 154, 170.

26. Aldo Leopold, *Game Management* (New York: Charles Scribner's Sons, 1933), 20.

27. Herbert L. Stoddard, *The Bobwhite Quail: Its Habits, Preservation, and Increase* (New York: Charles Scribner's Sons, 1931).

28. Herbert Stoddard, *Memoirs of a Naturalist* (Norman, Oklahoma: University of Oklahoma Press, 1969), 70–110. McAtee, head of the Survey's Division of Food Habits Research, pressed Nelson to make this appointment; see Box 80, Office Memos, Waldo L. McAtee Papers, Manuscript Division, Library of Congress.

29. Stoddard to Leopold, 5 December 1931, in Aldo Leopold Papers, Correspondence—Stoddard, University of Wisconsin Archives, Department of Wildlife Ecology, Madison, Wisconsin (hereafter cited as Leopold Papers, Madison, Wisconsin).

30. McAtee to J. N. Darling, Chief of the Survey, 27 April 1934, Box 80, Office Memos, McAtee Papers, Library of Congress.

31. The project continued under a number of graduate students. Errington summarized much of the work in "Some Contributions to a Fifteen-Year Local Study of the Northern Bobwhite to a Knowledge of Population Phenomena," *Ecological Monographs, 15* (January 1945), 3–34. There are more complete reports in Leopold

Papers, Madison, Wisconsin, under Research Areas and Projects—Prairie du Sac, and Correspondence—Errington.

32. Paul Errington, "Bobwhite Winter Survival in an Area Heavily Populated with Grey Foxes," *Iowa State College Journal of Science, 8* (1933–1934), 127–130. On Errington's ideas on this property of land, which he called at first "carrying capacity" and later "thresholds of security," see his correspondence with Leopold, 1940–1944, in Leopold Papers, Research Areas and Projects—Prairie du Sac, and Correspondence—Errington, Madison, Wisconsin. After the mid-1930s, the apparent stability of various areas within the study's boundaries broke down, and by the early 1940s there seemed to be no constant area smaller than the full five square miles. Nor, later work suggested, was the area as isolated as had been assumed, and the flow of birds across the boundaries complicated matters. Errington's ideas changed and developed. The main point—that predator control was not a cure-all—remained.

33. Errington discussed the confusion in *Of Predation and Life* (Ames, Iowa: Iowa State University Press, 1967). The phrase with underlined words comes from p. 235.

34. Paul Errington, "Vulnerability of Bob-white Populations to Predation," *Ecology, 15* (April 1934), 110–127; idem, "What is the Meaning of Predation?" in *Annual Report of the Smithsonian Institution for 1936* (Washington: Government Printing Office, 1937), 243–252; Errington and H. L. Stoddard, "Modifications in Predation Theory Suggested by Ecological Studies of the Bobwhite Quail," in *Transactions of the Third North American Wildlife Conference* (Washington: American Wildlife Institute, 1938), 736–740.

35. Leopold, *Game Management*, 252.

36. Olaus J. Murie, "Food Habits of the Coyote in Jackson Hole, Wyo.," USDA Circular 362, October 1935 (Washington: Government Printing Office, 1935); Adolph Murie, *Ecology of the Coyote in the Yellowstone*, Fauna Series Number Four, National Park Service (Washington: Government Printing Office, 1940).

37. Box 264, Miscellaneous P file, Murie Papers, Conservation Center, Denver Public Library, Colorado, has comments on Olaus's paper. See also Box 83, McAtee Papers, Library of Congress. On Adolph's troubles, see Olaus Murie to Harold E. Anthony, 5 December 1945, Murie file, Department of Mammalogy, American Museum of Natural History, New York, and Olaus Murie to Carl L. Hubbs, 30 July 1946, Correspondence, Box 361, Folder 7, Adolph Murie Papers, Denver Public Library. See also File 720, Entry 7, RG 79, Records of the National Park Service, National Archives.

38. Adolph Murie, *The Wolves of Mt. McKinley*, Fauna Series Number Five, National Park Service, Department of the Interior (Washington: Government Printing Office, 1944).

39. Murie's was the first study of large wild predators to consider ecology or ethology, and *The Wolves of Mt. McKinley* set as much a standard and an example in its area as Stoddard and Errington's work did in theirs. L. David Mech, an international authority on wolf biology, dedicated *The Wolf: The Ecology and Behavior of an Endangered Species* (Garden City, N.Y.: Natural History Press, 1970), to Murie for his pioneering work.

40. Murie, *Wolves of Mt. McKinley*, 24–25, 30–31, 26–27.

41. Olaus was not insensitive to these changes, and during the 1930s he was assimilating ecological ideas and research theories. See his letter to Marcus Ward

Lyon, 30 March 1935, on his work with Adolph on elk. Murie file, Department of Mammalogy, American Museum of Natural History, New York.

42. Leopold, *Game Management*, 3. The plan of the book also shows this parallel.

43. On conservation and efficiency, see Samuel P. Hays, *Conservation and the Gospel of Efficiency* (Cambridge: Harvard University Press, 1959).

44. Leopold Papers, Madison, Wisconsin, General Files, Department of Wildlife Ecology, 1933–1948, Box 1.

45. On the need for managers to be as adept at "selling" ideas as developing them, see University of Wisconsin Archives, General Files, Department of Wildlife Ecology, 1933–1948, 9/25/3, Box 1. On developing programs, see Rudolph Bennitt, "Nature and Scope of Training Graduates in Wildlife Conservation," and William Van Dersal, "The Viewpoint of Employers in the Field of Wildlife Conservation," in *Transactions of the Seventh North American Wildlife Conference, 1942* (Washington: American Wildlife Institute, 1942), 495–505, 506–517.

46. On the prevalence of this interpretation, see Caughley, "Eruption of Ungulate Populations," 54; Burk, "Kaibab Deer"; Wagner, "Livestock."

47. Allee et al., *Principles of Animal Ecology*, 706.

48. Reuben Edwin Trippensee, *Wildlife Management: Upland Game and General Principles* (New York: McGraw-Hill, 1948), 403.

49. Durward Allen, *Our Wildlife Legacy* (New York: Funk and Wagnalls, 1954), 236.

50. Aldo Leopold, "Game and Wild Life Conservation," *Condor, 34* (March 1932), 103–106, which is in reply to T. T. McCabe, *Condor, 33* (November 1931), 259–261. See also American Game Plan in "Report to the American Game Conference on an American Game Policy," in *Transactions of the Seventeenth American Game Conference* (no publisher, no date), held December 1930, 284–309; copy in University of Wisconsin Archives, General Files, Department of Wildlife Ecology, 1933–1948, Box 1.

51. Office Memos, 1934–1938, McAtee Papers, Library of Congress.

52. W. L. McAtee, "Long Live the Wildlife Society," *Journal of Wildlife Management, 1* (July 1937), 45–46; idem, "Notes," *Journal of Wildlife Management, 2* (April 1938), 61; Victor H. Cahalane, "Report of the Secretary," *Journal of Wildlife Management, 2* (April 1938), 65–68.

53. On McAtee's division and its work on game species, see his letter to J. N. Darling, 27 April 1934, Box 80; on Survey and state universities, see Office Memos, 1934; on Survey and Wildlife Society, see Office Memos 1937 and 1938, Box 80. All in McAtee Papers, Library of Congress.

54. See "Report of the United States Biological Survey" in U.S. Congress, Senate, Special Committee on the Conservation of Wildlife Resources, *The Status of Wildlife in the United States*, Senate Report 1203, 76th Congress, 3rd Session (Washington: Government Printing Office, 1940), 65–160; Flader, *Thinking Like a Mountain*, 26–28; Office Memos, 1934–1938, Box 80, McAtee Papers, Library of Congress.

55. A. Murie, *Ecology of the Coyote in the Yellowstone*, 8–16. See also Victor H. Cahalane, "The Evolution of Predator Control Policy in the National Parks," *Journal of Wildlife Management, 4* (July 1939), 229–237. The following account draws heavily on material in Files 719 and 720, Entry 7, Central Classified File, 1933–1949, Records of the National Park Service, RG 79, National Archives.

56. Horace M. Albright, "The National Park Service's Policy on Predatory Mammals," *Journal of Mammalogy, 12* (May 1931), 185–186.

57. Samuel J. Harbo, Jr., and Frederick C. Dean, "Historical and Current Perspectives on Wolf Management in Alaska," in Ludwig N. Carbyn (editor), *Wolves in*

Canada and Alaska (Ottawa: Canadian Wildlife Service, 1983), 52; Cahalane, "Predator Control in the Parks."

58. File 720, Entry 7, RG 79, National Archives. See also George M. Wright, Joseph S. Dixon, Ben H. Thompson, *Fauna of the National Parks of the United States*, Fauna Series Number One, Department of the Interior (Washington: Government Printing Office, 1933); "Report of the U.S. National Park Service," in Senate, *Status of Wildlife*, 350–352; Cahalane, "Evolution of Predator Control Policy."

59. Wright et al., *Fauna of the National Parks*, 54.

60. Wright et al., *Fauna of the National Parks*, 2.

61. Wright et al., *Fauna of the National Parks*, 54.

62. L. David Mech, *The Wolves of Isle Royale*, Fauna of the National Parks of the United States, Number Seven (Washington: Government Printing Office, 1966); Isle Royale file, Entry 34, RG 79, National Archives.

63. Senate, *Status of Wildlife*, 354–365. On endangered wildlife, a topic that became important in the 1930s, see United States Department of the Interior, *Fading Trails: The Story of Endangered American Wildlife* (New York: Macmillan, 1943).

64. Harold B. Bryant and Newton B. Drury, "Development of the Naturalist Program in the National Park Service," an interview conducted by Amelia R. Fry, Regional History Office, University of California, Berkeley, 1964, courtesy of The Bancroft Library; Cahalane, "Evolution of Predator Control Policy."

65. File 720, Entry 7, RG 79, National Archives.

66. Senate, *Status of Wildlife*, 355.

67. A review of this work is in Senate, *Status of Wildlife*.

Chapter Six. From Knowing to Feeling

1. Keith Thomas, *Man and the Natural World* (New York: Pantheon, 1983); James C. Turner, *Reckoning with the Beast* (Baltimore: Johns Hopkins University Press, 1980).

2. Alfred Runte, *The National Parks: The American Experience* (Lincoln: University of Nebraska Press, 1979); Runte, personal communication.

3. Frank M. Chapman, "A Note on the Economic Value of Gulls," *Bird-Lore*, 2 (February 1900), 10.

4. Editorial by Frank M. Chapman, *Bird-Lore*, 22 (January-February 1920), 53; A. K. Fisher, *Hawks and Owls of the United States in Relation to Agriculture*, Bulletin Three, Bureau of Biological Survey (Washington: Government Printing Office, 1892), was a favorite text. On Audubon's defense of birds, see *Bird-Lore*, from 1899.

5. See, for example, Mabel Osgood Wright, "Stories from a Bird Sanctuary, II, The Rights and Wrongs of Bird-trapping," *Bird-Lore*, 24 (September-October 1922), 253–255, and the arguments in 1909–1910 for and against shooting house wrens, which were bullies and drove away more desirable birds.

6. For examples of the Society's attitudes, see editorial, *Bird-Lore*, 32 (November-December 1930), 446; "Campaign for Hawk and Owl Protection," Resolution of the Directors of the National Association of Audubon Societies, 9 May 1934, *Bird-Lore*, 36 (September-October 1934), 333–335.

7. "Feathered vs. Human Predators," *Bird-Lore*, 37 (March-April 1935), 122–126; W. L. McAtee, "A Little Essay on 'Vermin,' " *Bird-Lore*, 33 (November-December 1931), 381–384.

8. Warren F. Eaton, "Predators and Bird Preserves," *Bird-Lore*, 37 (May-June

1935), 162–166. Compare this with *Bird-Lore, 32* (November-December 1930), 446, or Wright, "Stories from a Bird Sanctuary," 254–255.

9. Aldo Leopold, "The State of the Profession," *Journal of Wildlife Management, 4* (July 1940), 344; Donald Culross Peattie, *A Prairie Grove* (New York: Literary Guild, 1938). In 1944 Edwin Way Teale included three of Peattie's books in an article on "The Great Companions of Nature Literature," a selection of one hundred great nature volumes; see Edwin Way Teale, "The Great Companions of Nature Literature," *Audubon, 46* (November-December 1944), 363–366.

10. Peattie, *A Prairie Grove,* 6.

11. Peattie, *A Prairie Grove,* 6, 19.

12. Peattie, *A Prairie Grove,* 84.

13. Peattie, *A Prairie Grove,* 284.

14. Rachel Carson, *Under the Sea Wind* (Boston: Houghton Mifflin, 1941).

15. Henry Beston, *The Outermost House* (Garden City, N.Y.: Doubleday, 1928).

16. Carson, *Under the Sea Wind,* xiii–xiv.

17. Sally Carrighar, *One Day at Beetle Rock* (New York: Alfred Knopf, 1944).

18. John F. Stanwell-Fletcher, "Three Years in the Wolves' Wilderness," *Natural History, 49* (March 1942), 137–147. See also Theodore C. Stanwell-Fletcher, *Driftwood Valley* (London: George G. Harrap, 1949), and "The Wolves at Tetana," *Atlantic, 178* (August 1946), 103–109.

19. On this theme and wolves, see a somewhat later work, Lois Crisler, *Arctic Wild* (New York: Harper, 1958).

20. Victor H. Cahalane, "Shall We Save the Larger Carnivores?" *The Living Wilderness, 17* (1946), 17–22.

21. Aldo Leopold, *A Sand County Almanac* (New York: Oxford University Press, 1949; reprinted New York: Ballantine Books, 1970). I, like others who have worked on Leopold, rely heavily on the work of Susan Flader. Her *Thinking Like a Mountain* (Columbia, Missouri: University of Missouri Press, 1974) is the basic guide. Unless otherwise footnoted, my information on Leopold comes from her, and the description of Leopold's changing ideas, chronology, and major influences has been formed by my study of her work. Flader arranged the Leopold papers in the University of Wisconsin Archives; this was an exceptionally well-done job and has been a great aid in my own research in Leopold's papers.

22. Flader, *Thinking Like a Mountain.*

23. Leopold, *Sand County Almanac,* 138.

24. Unpublished manuscript, headed FORWARD (Revision of 7/31/47) (to be revised as appendix), from University of Wisconsin Archives, Department of Wildlife Ecology, Madison, Wisconsin, Box 17, Unpublished Manuscripts—Typescript Copies. These papers hereafter cited as Leopold Papers, Madison, Wisconsin. With permission of A. Starker Leopold.

25. Incident described in New Mexico Journal, 1917–1924, in Diaries and Journals, Series 9/25/10–7, Box 2, Leopold Papers, Madison Wisconsin. Paper is in "Species and Subjects, Big Game: Upland Game Birds," Series 9/25/10–4, Box 2, Leopold Papers. Swarth to Leopold, 19 July 1922, Leopold to Swarth, 21 July 1922, Leopold file, Correspondence of the Museum of Vertebrate Zoology, Berkeley. With permission of David Wake, Director. Starker Leopold described his father's reaction in an interview with the author, 16 June 1981. At some unknown time, Leopold inserted in his own files (Species and Subjects: Predators and Furbearers Box) a quote from *Game Birds of California* (1918) by Joseph Grinnell, Harold Bryant, and Tracy Storer (all of them with the Museum of Vertebrate Zoology)—a

quote to the effect that the accusations against roadrunners on the grounds that they ate young quail were exaggerated and "the killing of this bird as an injurious species is wholly unjustified."

26. Leopold, Unfinished Manuscripts, FORWARD, Box 17, Leopold Papers, Madison, Wisconsin.

27. Unpublished manuscripts, Box 17, Leopold Papers, Madison, Wisconsin.

28. The survey covered Minnesota, Iowa, Missouri, Illinois, Wisconsin, Indiana, Michigan, and Ohio. Since the area was almost entirely farm land, the survey was focused on the problems of game production on land intended for other uses as well, which involved Leopold in the problems of balancing interests and species, problems which played an important role in his thinking about the "land ethic." The report is *Report on a Game Survey of The North Central States* (Madison, Wisconsin: Sporting Arms and Ammunition Manufacturers Institute, 1931).

29. Flader, *Thinking Like a Mountain*, 24. See also Stoddard correspondence in Leopold Papers, Madison, Wisconsin.

30. Aldo Leopold, *Game Management* (New York: Charles Scribner's Sons, 1933), Chapter 10. The Prairie du Sac Project, which lasted from 1929 to 1945, is particularly important; see notes in Prairie du Sac Project and Correspondence—Errington, in Leopold Papers, Madison, Wisconsin. It includes Leopold's correspondence with Errington from 1940 to 1944 on an article on the project's significant results (the collaboration eventually fell through). Applied ecological work was a major source of information for the "pure" scientists. In 1954, discussing population regulation in birds, the English ecologist David Lack noted that predation "has been studied chiefly in the United States of America, stimulated by the commercial incentive of game management"; see David Lack, *The Natural Regulation of Animal Numbers* (London: Oxford University Press, 1954), 154.

31. Flader, *Thinking Like a Mountain*, 139–154.

32. Flader, *Thinking Like a Mountain*, Chapters 5 and 6.

33. Leopold, *Sand County Almanac*, 262.

34. Leopold to Richard Bond, Soil Conservation Service, Berkeley, California, 23 November 1939, Predator Control—Predation, Subject files, Leopold Papers, Madison, Wisconsin.

35. Aldo Leopold, "Wolves," in Unpublished Manuscripts—typescript copies, Box 17, 9/25/10–6, Leopold Papers, Madison, Wisconsin.

36. Leopold, *Sand County Almanac*, 137–141.

37. On the early humane movement, see James Turner, *Reckoning with the Beast* (Baltimore: Johns Hopkins University Press, 1980); Lisa Mighetto, "Wild Animals in American Thought and Culture, 1870s–1930s" (Ph.D. dissertation, University of Washington, 1986), has the best study of the full range of nature stories. See also idem, "Science, Sentiment, and Anxiety: American Nature Writing at the Turn of the Century," *Pacific Historical Review, 54* (February 1985), 33–50. I am also indebted to Mighetto for personal communications. See also Robin Doughty, *Feather Fashions and Bird Preservation* (Berkeley: University of California Press, 1975). On the anti-trap movement, see John R. Gentile, "The Evolution and Geographic Aspects of the Anti-trapping Movement: A Classic Resource Conflict" (Ph.D. dissertation, Oregon State University, 1984).

38. Gentile, "The Anti-Trapping Movement," 67; Bailey Papers, Smithsonian Institution Archives, Box 5.

39. Gentile, "Evolution of Anti-Trapping Movement," 61–66.

40. Edward Breck, *The Steel-Trap: A Manual of Information* (Washington: Anti-

Steel Trap League, Inc., n.d.), 28, cited in Gentile, "The Anti-Trapping Movement," 61.

41. Theodore W. Cart, "The Struggle for Wildlife Protection in the United States, 1870–1900: Attitudes and Events Leading to the Lacy Act" (Ph.D. dissertation, University of North Carolina, 1971), 115. Poem in Edward Breck, "Blood Money for the Audubon Association," 2, credited to F. F. Van Water; copy in Folder 24, Box 291, Rosalie Edge Papers, Conservation Center, Denver Public Library, Colorado. It later appeared on the back of the stationery of Defenders of Wildlife (1963); example in Olaus J. Murie Papers, Denver Public Library, Box 265, "Predators—Coyotes" file. See also Gentile, "The Anti-Trapping Movement," 62.

42. Bailey combined a lively interest in humane traps—he invented a cage trap for beaver in 1925 as well as the Verbail trap—with a commitment to the Survey's poisoning program, an indication of the complicated attitudes and ideas which people might have about nature and animals. See Bailey Papers, Smithsonian Institution Archives, Box 5.

43. Lucy Furman, "The Price of Furs: A Plea for Humane Trapping," *Atlantic Monthly, 141* (February 1928), 206–209; Tom Wallace, "Ninety Pounds of Fight," *Nature Magazine, 35* (February 1942), 95–96.

44. The following account of Edge's work is drawn from her papers in the Conservation Center, Denver Public Library.

45. Pamphlets and correspondence in Edge Papers, Denver Public Library.

46. Emergency Conservation Committee, "The United States Bureau of Destruction and Extermination: The Misnamed and Perverted 'Biological Survey,' " Edge Papers, Denver Public Library.

47. F. F. Van Water, copy in Folder 24, Box 291, Edge Papers, Denver Public Library.

48. Edge Papers, Denver Public Library; W. L. McAtee to Joseph Grinnell, 7 July 1931, McAtee file, Museum of Vertebrate Zoology, Berkeley.

49. T. Gilbert Pearson, *Adventures in Bird Protection* (New York: Appleton-Century, 1937), 215–218; Joseph Grinnell, "Conserve the Collector," *Science, 41* n.s. (February 1915), 229–232. It appears in a posthumous collection of essays, *Joseph Grinnell's Philosophy of Nature* (1943; reprinted Freeport, N.Y.: Books for Libraries Press, 1968), 65–72. Storer Correspondence, Joseph Grinnell Papers, Bancroft Library, University of California, Berkeley.

50. Edge to Williams, 6 December 1938, Folder 17, Box 291, Edge Papers, Denver Public Library.

Chapter Seven. The Public and Ecology

1. Alan Devoe, "Why Bother?" *Audubon, 50* (November-December 1948), 332–335.

2. Fairfield Osborn, *Our Plundered Planet* (Boston: Little, Brown, 1948); William Vogt, *Road To Survival* (New York: William Sloane, 1948).

3. Osborn, *Our Plundered Planet*, viii.

4. Osborn, *Our Plundered Planet*, frontispiece.

5. Leopold to Vogt, Correspondence files, Starker Leopold File, University of Wisconsin Archives, Department of Wildlife Ecology, Madison, Wisconsin.

6. Robert Boardman, *International Organization and the Conservation of Nature* (Bloomington, Indiana: Indiana University Press, 1981), 35–46.

7. The interest in wilderness can best be traced in the postwar period through

the annual conferences. For a sampling see William Schwartz (editor), *Voices for the Wilderness* (New York: Ballantine, 1969).

8. Roderick Nash, *Wilderness and the American Mind* (New Haven: Yale University Press, 1983), 220–222. Chapter 12 describes the rise of wilderness sentiment in the postwar years. For more detailed studies of aspects of this question, see Richard A. Baker, "The Conservation Congress of Anderson and Aspinall, 1963–64," *Journal of Forest History, 29* (July 1985), 104–119; Dennis M. Roth, *The Wilderness Movement and the National Forests: 1964–1980* (Washington: Government Printing Office, 1984). A study of changes in public ideas is in Linda Graber, *Wilderness as Sacred Space* (Washington: Association of American Geographers, 1976).

9. The following account of DDT's career is based on Thomas R. Dunlap, *DDT: Scientists, Citizens, and Public Policy* (Princeton: Princeton University Press, 1981).

10. Rachel Carson, *Silent Spring* (Boston: Houghton Mifflin, 1962).

11. Carson, *Silent Spring*, 18, 22.

12. O. J. Murie to Struthers Burt, 5 December 1945, Box 361, Correspondence files, Olaus Murie Papers, Conservation Center, Denver Public Library, Colorado.

13. Graber, *Wilderness as Sacred Space*; Nash, *Wilderness and the American Mind*, Chapters 12 and 13. On national parks, see Alfred Runte, *National Parks: The American Experience* (Lincoln, Nebraska: University of Nebraska Press, 1979).

14. John F. Stanwell-Fletcher, "Three Years in the Wolves' Wilderness," *Natural History, 49* (March 1942), 137–147. See also Theodore C. Stanwell-Fletcher, *Driftwood Valley* (London: George G. Harrap, 1949), and "The Wolves at Tetana," *Atlantic, 178* (August 1946), 103–109.

15. Lois Crisler, *Arctic Wild* (New York: Harper, 1958). Quote is from 285.

16. Its story is in Stanley Paul Young, *The Last of the Loners* (New York: Macmillan, 1970). See also "History of Predator Control—General," in General Records, Division of Wildlife Services, Record Group 22, Records of the Fish and Wildlife Service, National Archives, Washington, D.C.; W. B. Bell, "Hunting Down Stock Killers," in United States Department of Agriculture, *Yearbook of the Department of Agriculture, 1920* (Washington: Government Printing Office, 1921), 289–300.

17. Roger Caras, *The Custer Wolf* (Boston: Little, Brown, 1966), 25, 30, 59, 175.

18. Caras, *The Custer Wolf*, 171, 175.

19. Farley Mowat, *Never Cry Wolf* (Boston: Little, Brown, 1963).

20. Whether Mowat gave an accurate account of his own experiences has been debated. Some reviewers, including his supervisor, have claimed that the book is fiction; see A.W.F. Banfield, *Canadian Field Naturalist, 78* (1964), 52–54, and Douglas H. Pimlott, *Journal of Wildlife Management, 30* (1964), 236. Canadian Wildlife Service officials were amused and outraged. They saw it as a fictitious narrative that cast aspersions on the Service; see Wolves, WLU 200, Box 77, Record Group 109, Records of the Canadian Wildlife Service, Public Archives Canada, Ottawa. This is not our concern. The natural history—that is, the observations—are congruent with those others have made, and the book's influence, our main consideration, would not be affected even if they were completely made up.

21. Mowat, *Never Cry Wolf*, 162–163, 164.

22. Ernest Thompson Seton, *Wild Animals I Have Known* (New York: Charles Scribner's Sons, 1898), 41.

23. Barry Holstun Lopez, *Of Wolves and Men* (New York: Charles Scribner's Sons, 1978).

24. R. D. Lawrence, *In Praise of Wolves* (New York: Henry Holt, 1986).

25. Dan Strickland, "Wolf Howling in Parks—the Algonquin Experience in Interpretation," in Ludwig N. Carbyn (editor), *Wolves in Canada and Alaska*, Canadian Wildlife Service Report, Series Number 45 (Ottawa: Canadian Wildlife Service, 1983), 93–95.

26. Alexis de Tocqueville, *Democracy in America*, Volume I (1832; New York: Random House, 1945), 198.

27. The National Wildlife Federation's annual *Conservation Directory* is not a complete listing of organizations, but it provides a rough measure of organizational activity. Since it gives dates when organizations were formed, it can be shown that increased listings are not a reflection of better reporting but an actual increase in numbers. The following analysis is based on crude samples—every tenth organization listed in the nongovernmental category (International, National, Interstate Organizations, Commissions)—from incomplete surveys of back issues. No claim is made except that these figures indicate a large change in the number of organizations and a movement toward specialization.

28. *Conservation Directory*, 1983, lists some 40 organizations, of 392, that are identifiably single-purpose. Fifteen of these are for game species, six for raptors, five for sea mammals (three for whales specifically), three for carnivorous mammals (two for wolves, one for bear), and the rest scattered. The 1986 edition shows much the same picture.

29. Quantitative measures of public participation can most easily be found in a series of reports by Stephen R. Kellert and others. They are Stephen R. Kellert, *Public Attitudes Toward Critical Wildlife and Natural Habitat Issues* (Washington: Government Printing Office, 1979); idem, *Activities of the American Public Relating to Animals* (Washington: Government Printing Office, 1980); Kellert and Joyce Berry, *Knowledge, Affection and Basic Attitudes Toward Animals in American Society* (Washington: Government Printing Office, 1980); Kellert and Miriam O. Westervelt, *Trends in Animal Use and Perception in Twentieth Century America* (Washington: Government Printing Office, 1981).

Chapter Eight. Poisons and Policy

1. Ten-eighty was much more toxic to canids and rodents than to other forms of life, a property that made it attractive to the PARC. Eric Peacock, "Sodium Monofluoroacetate" (1964), typescript in Correspondence 1080, General Records, Division of Wildlife Services, Records of the United States Fish and Wildlife Service, Record Group 22, National Archives, Washington, D.C. (hereafter cited as RG 22, National Archives). On postwar condor poisoning incidents, see Report of District Agent to Assistant Director re Cal. Condor Incident from Kern County, March, 1950, copy in Gottschalk Correspondence, Museum of Vertebrate Zoology, Berkeley, California, with permission of David Wake, Director (hereafter cited as MVZ). See also Advisory Board, Predators, MVZ.

2. There is a large collection of literature in Poison 1080, Articles and Publications, 1945–1968, General Files, Division of Wildlife Research, RG 22, National Archives, relating to this issue. For a strong public statement of the case against the program, see Jack Olsen's *Slaughter the Animals, Poison the Earth* (New York: Simon and Schuster, 1971).

3. Poison 1080 files in General Files, Division of Wildlife Research, RG 22, National Archives.

4. See various "Poison" files, "History of Predator and Rodent Control, General," and "Coyote Getter," in General Files, Division of Wildlife Services, RG 22, National Archives. Traps (of which the PARC received many samples) remained, and aerial gunning was tested, but the emphasis was on poison.

5. Harold Anthony to Grinnell, 19 October 1931, Anthony file, MVZ.

6. Weldon B. Robinson, "Merits and Demerits of Thallium in the Control of Coyotes," 28, in Research Reports, Division of Wildlife Research, RG 22, National Archives.

7. Robinson, "Merits," 44–46, 36–37. See also letter of E. R. Kalmbach in this file.

8. E. R. Kalmbach to Ira N. Gabrielson, 2 September 1944, with Robinson's 1944 manuscript, "The Thallium-Studded Station as a Means of Coyote Control in Acute Predation Areas," in Research Reports, Division of Wildlife Research, RG 22, National Archives.

9. Lewis Laney, "New War Born 1080 Coyote Poison May Kill All Predators," *New Mexico Stockman* (May 1948), 75. Copy in 1080 Articles and Publications, General Records, Division of Wildlife Services, RG 22, National Archives.

10. Peacock, "Sodium Monofluoroacetate." Memo 121, 24 October 1945, by Dorr Green, in Correspondence 1080, General Records, Division of Wildlife Services, RG 22, National Archives. See also E. R. Kalmbach, Wildlife Research Laboratory, Denver, to Lloyd W. Swift, Division of Wildlife Management, Forest Service, 8 May 1947. Copy in Advisory Board on Wildlife Management—Predator Control, 1964, MVZ.

11. Kalmbach to Swift, 8 May 1947, copy in Advisory Board on Wildlife Management—Predator Control, 1964, MVZ. Material in the Aldo Leopold files for 1935 (MVZ) indicates, though, that condors do not regurgitate. In 1950 one condor was killed and another possibly killed at a strychnine bait station in Kern County, California, but PARC officials discounted the poison. See Gottschalk file, MVZ.

12. "Statement of Policy Adopted by Fish and Wildlife Service for Use of Compound 1080 (Sodium Monofluoroacetate) in Poison Stations to Kill Coyotes" (5 November 1947), in 1080—Misc. 1946–1952, ADC, General Files, Division of Wildlife Services, RG 22, National Archives. The set of memos in this file, though incomplete, is useful in tracing the internal debate over use of the new chemical.

13. E. R. Kalmbach to C. C. Presnall, 14 February 1949, in 1080 Correspondence—Instructions to Regions (ADC), General Records, Division of Wildlife Services, RG 22, National Archives.

14. E. R. Kalmbach, memo of 10 January 1951, in Poison 1080—Studies of, in General Records, Division of Wildlife Services, RG 22, National Archives. The situation, though, did not change. In the late 1960s the Service still placed about 15,000 bait stations a year; memo of 16 September 1969 in Poison 1080—Studies of, General Records, Division of Wildlife Services, RG 22, National Archives.

15. Cottam to Day, 31 October 1951, Poison 1080, Complaints, in General Records, Division of Wildlife Services, RG 22, National Archives.

16. Poison 1080—Violations, in General Files, Division of Wildlife Services, RG 22, National Archives.

17. John Gottschalk to Stanley Cain, 19 October 1965, restricted memo, in Poison 1080, 1965–1966, General Files, Division of Wildlife Services, RG 22, National Archives. Other files under 1080 and Poison 1080, including correspondence on use, furnish additional information.

18. Cottam to Leopold, 8 February 1956, Cottam file, MVZ. See also Advisory Board on Wildlife Management, 1963, MVZ.

19. *The National Woolgrower* provides an excellent running account of the economic pressures on the industry in the postwar period.

20. Charles L. Cadieux to Howard J. Matley, 16 January 1961, memo on proposed article on 1080, in Poison 1080, 1960–1963, General Files, Division of Wildlife Services, RG 22, National Archives.

21. None of this was new. Trappers had backed Dixon's investigation of the furbearers of Nevada in 1926, part of the evidence used against the Survey; hunters had worried about some of the PARC's actions for years; and bait stations, whatever the chemical used, had always posed a danger to dogs. See Poison; Use of Poisons, Criticism of, 1949–1965; Poison 1080, Criticism of, Use of, 1947–1964. On early dog poisoning, see "Does Coyote Poisoning Work on the High Ranges Defeat Our Purpose in Mountain Lion Eradication?" by M. E. Musgrave, Conference on Predator and Rodent Control, Ogden, Utah, 23–28 April 1928, Reports of Conferences, 1928–1941. All material in General Files, Division of Wildlife Services, RG 22, National Archives.

22. Melvin Smith to C. C. Presnall, 6 December 1957 (personal), Everett M. Mercer to Regional Director, 29 September 1959 in Poison 1080, General Files, Division of Wildlife Services, RG 22, National Archives.

23. Walter E. Howard, "1080—A Rodent Poison of Controversy," *California Farmer*, 17 September and 15 October 1960.

24. Faith McNulty, *The Whooping Crane* (New York: E. P. Dutton, 1966); Robert Allen, *The Whooping Crane*, Research Report Number Three, National Audubon Society (New York: National Audubon Society, 1952). See also Whooping Crane files, WLU 61, Records of the Canadian Wildlife Service, Record Group 109, Public Archives Canada, Ottawa.

25. In 1983 and 1984 scientists, using radiotracking, found almost one hundred ferrets in northwestern Wyoming. These animals recently suffered from an epidemic of distemper, and there are fears that the species is now gone from the wild. Ferret files, Office of Endangered Species, U.S. Fish and Wildlife Service, Washington, D.C.

26. Stanley Paul Young, "Black Boots of the Prairie," *American Forests, 46* (January 1940), 16–18; Elliott Coues, *The Fur-Bearing Animals of North America*, U.S. Department of the Interior, Geological Survey, Miscellaneous Publication Number Eight (Washington: Government Printing Office, 1877; reprinted New York: Arno, 1970), 149–153.

27. Coues, *Fur-Bearing Mammals*, 153.

28. Young, "Black Boots," 18; Ernest Thompson Seton, *Lives of Game Animals*, Volume II, Part II (1900, 1926; reprinted Boston: Charles Branford, 1953), 565–574. Young to Buell, 10 December 1953, Folder 2, Box 2, Stanley Paul Young Papers, Smithsonian Institution Archives, Washington, D.C.

29. Seton, *Lives of Game Animals*, 573–574.

30. Green to Farley, 11 February 1954, and Farley to Regional Directors, 13 April 1954, Ferret—Correspondence, 1950–1965, General Files, Division of Wildlife Services, RG 22, National Archives.

31. Nobel Buell to the South Dakota Department of Game, Fish, and Parks, 22 April 1954, in Ferret—Correspondence, 1950–1965, General Files, Division of Wildlife Services, RG 22, National Archives.

32. Memo of 14 November 1960, in Ferret—Correspondence, 1950–1965, RG 22, National Archives.

33. Clifford C. Presnall, 26 August 1963, in Ferret file, R.G. 22, National Archives.

34. On early history, see Endangered Species notebook, Office of Endangered Species, Fish and Wildlife Service. Courtesy of Ronald Nowak.

35. Memo of 18 May 1965, in Ferret file, General Files, Division of Wildlife Services, RG 22, National Archives.

36. Carbon copy of undated letter, Assistant. Director—Wildlife, to Regional Director, Minnesota (possibly never sent). Position in file and references to Pine Ridge indicate early 1965; see Ferret file, RG 22, National Archives.

37. Ferret file, General Files, Division of Wildlife Services, RG 22, National Archives. See also Faith McNulty, *Must They Die?* (Garden City, N.Y.: Doubleday, 1971). This first appeared as "Prairie Dogs and Black-Footed Ferrets," *The New Yorker*, 13 June 1970, 40.

Chapter Nine. Ending the Poisoning

1. Gabrielson gave Eugene Kellogg information and opinions that Kellogg had used to attack Grinnell and Linsdale on the issue of thallium-poisoned grain in California—which led to a stiff exchange of letters with Grinnell. See Gabrielson file, Museum of Vertebrate Zoology, Berkeley, California, with permission of David Wake, Director (hereafter MVZ). See Gabrielson to Witmer Stone, president of American Society of Mammalogists, in Anthony file, MVZ. See address to 1936 Conference on Predator and Rodent Control, in Conference file, 1928–1941, General Files, Division of Wildlife Services, Records of the United States Fish and Wildlife Service, Record Group 22, National Archives, Washington, D.C. (hereafter RG 22, National Archives).

2. Ira N. Gabrielson, "Must the Antelope Go?" *American Forests, 41* (October 1935), 575–576; idem, "What Can We Do About Our Rare and Vanishing Species?" *Scientific American* (January 1938), 5–8; idem, *Wildlife Conservation* (New York: Macmillan, 1941), 194–210.

3. Advisory Board on Wildlife Management, "Predator and Rodent Control in the United States," *Transactions of the Twenty-ninth North American Wildlife and Natural Resources Conference* (Washington: Government Printing Office, 1964), 6, 15.

4. Advisory Board, "Predator and Rodent Control," 13.

5. Jack Berryman, author's interview, 17 December 1979; John Gottschalk, author's interview, 20 December 1979.

6. "Man and Wildlife," 4, 5, 10. Copy in "Policy—Corresp. re Policy, Man and Wildlife," in General Files, Division of Wildlife Services, RG 22, National Archives.

7. "Policy—Corresp. re Policy, Man and Wildlife" file, in General Files, Division of Wildlife Services, RG 22, National Archives.

8. Dingell was not looking for publicity or attempting to curry favor with his constituents—who were mainly working-class Democrats living near Ford's River Rouge plant. He was an ardent duck hunter and outdoorsman, interested in wetlands, wildlife, and wildlife habitat long before they were public issues.

9. U.S. Congress, House of Representatives, Subcommittee on Fisheries and Wildlife Conservation of the Committee on Merchant Marine and Fisheries, *Predatory Mammals*, Hearings before the Subcommittee on Fisheries and Wildlife Conservation of the Committee on Merchant Marine and Fisheries, 89th Congress, 2nd Session (Washington: Government Printing Office, 1966), 1, 154–157, 214–239.

10. U.S. Congress, House, *Predatory Mammals* (1966), 1.

11. *Defenders*, early issues, incomplete, in Library of Congress. On roots of this movement, see James Turner, *Reckoning with the Beast* (Baltimore: Johns Hopkins University Press, 1981). On trapping, see John R. Gentile, "The Evolution and Geographic Aspects of the Anti-Trapping Movement: A Classic Resource Conflict" (Ph.D. dissertation, Oregon State University, 1984).

12. Peter Singer, *Animal Liberation: A New Ethic for Our Treatment of Animals* (New York: Random House, 1975), 9, 18.

13. See, for example, Tom Regan and Peter Singer (editors), *Animal Rights and Human Obligations* (Englewood Cliffs, N.J.: Prentice-Hall, 1976), or Peter Singer (editor), *In Defense of Animals* (Oxford: Basil Blackwell, 1985). Environmentalists raised the related, but distinct, question of legal rights. The most famous exposition of this was Christopher D. Stone, *Should Trees Have Standing?* originally published in *Southern California Law Review*, 1972. It is available with excerpts from Supreme Court cases that used it, as a book (Los Altos, California: William Kaufmann, 1974).

14. Singer, *Animal Liberation*, x, 7, 18. Briton Cooper Busch provides a study of the clash of animal rights activists and wildlife managers over the killing of baby harp seals in *The War Against the Seals* (Montreal: McGill-Queen's University Press, 1985), Chapter 8.

15. Officers of conservation organizations, using a random sample from the National Wildlife Federation's annual *Conservation Directory*, 1975 and 1983 editions, were about 80 percent male. Organizations listing humane goals as part of their program had approximately equal numbers of men and women as officers, and no particular pattern of office holding. Hunting groups, by comparison, had almost entirely male officers. On distribution of ideas about nature and membership in wildlife protection organizations, see Stephen R. Kellert, *Activities of the American Public Relating to Animals* (Washington: Government Printing Office, 1980), 133–136.

16. Stephen R. Kellert and Joyce Berry, *Knowledge, Affection, and Basic Attitudes Toward Animals in American Society* (Washington: Government Printing Office, 1980).

17. Michael Bean, *The Evolution of National Wildlife Law* (revised edition; New York: Praeger, 1983), 98–103. On anti-hunting, see such works as Cleveland Amory, *Man Kind? Our Incredible War on Wildlife* (New York: Harper and Row, 1974), and Lewis Regenstein, *The Politics of Extinction: The Shocking Story of the World's Endangered Wildlife* (New York: Macmillan, 1975). The opposing view may be found in *Field and Stream, Outdoor Life, Guns and Ammo*, and *American Rifleman*.

18. Comments from professional wildlife advocates on a paper I gave ("The Federal Government, Wildlife, and Endangered Species," Woodrow Wilson Center, Smithsonian Institution, Conference on the Evolution of American Environmental Politics, 28, 29 June 1984) stressed this issue of incompatible goals. It was a very sore point.

19. Faith McNulty, "A Reporter At Large: Prairie Dogs and Black-footed Ferrets," *The New Yorker* (13 June 1970), 40, and *Must They Die?* (Garden City, N.Y.: Doubleday, 1971).

20. Jack Olsen, "Poisoning of the West," *Sports Illustrated, 34* (7 March 1971), 80–93; *34* (15 March 1971), 36–44; *34* (22 March 1971) 34–42; idem, *Slaughter the Animals, Poison the Earth* (New York: Simon and Schuster, 1971).

21. U.S. Congress, Senate, Subcommittee on Agriculture, Environmental and Consumer Protection of the Committee on Appropriations, *Predator Control and Related Problems*, Hearings before the Subcommittee on Agriculture, Environmental

and Consumer Protection of the Committee on Appropriations, 92nd Congress, 1st Session (Washington: Government Printing Office, 1971), 4–8, 11, 32.

22. U.S. Congress, Senate, *Predator Control and Related Problems* (1971), 40–41, 145–167. This appeared in the *Wall Street Journal* under the title, "Coyote Killing: Business as Usual," on 14 December 1971. U.S. Senate, *Predator Control and Related Problems* (1971), 10–13.

23. U.S. Congress, Senate, *Predator Control and Related Problems* (1971), 2nd Session, 78–79.

24. U.S. Congress, Senate, *Predator Control and Related Problems* (1971), 2nd Session.

25. Amory's *Man Kind?* cited above, is a good example of his stands.

26. Angus A. MacIntyre, "The Politics of Nonincremental Domestic Change: Major Reform in Federal Pesticide and Predator Control Policy" (Ph.D. dissertation, University of California, Davis, 1982), Chapter 6. Personal communication with McIntyre and conversation with various biologists. None of them spoke for the record, nor were they closely involved. No claim is made of complete knowledge.

27. The others were John A. Kadlec, Durward L. Allen, Richard A. Cooley, Maurice G. Hornocker, and Frederick H. Wagner.

28. Advisory Committee on Predator Control, *Predator Control—1971* (Ann Arbor: Institute for Environmental Quality, 1971), 2–3. A somewhat different assessment is in National Research Council, Subcommittee on Vertebrate Pests, Committee on Plant and Animal Pests, Agricultural Board, *Principles of Plant and Animal Control*, Volume 5: *Vertebrate Pests: Problems and Control* (Washington: National Academy of Sciences, 1970).

29. Advisory Board, *Predator Control—1971*, 3.

30. For Cain's comments on how the recommendations were carried out, see Stanley A. Cain, "Predator and Pest Control," in Howard P. Brokaw (editor), *Wildlife in America* (Washington: Government Printing Office, 1978), 379–395.

31. MacIntyre, "Nonincremental Change," provides the most thorough and balanced presentation of this view.

32. Jack Berryman indicated exactly this in an interview with the author, 17 December 1979.

33. Graeme Caughley, *The Deer Wars* (Auckland: Heinemann, 1983), 119.

34. MacIntyre, "Nonincremental Change," Chapter 6.

35. MacIntyre, "Nonincremental Change," Chapter 6.

36. Starker Leopold, author's interview, 18 June 1981, suggested that Nixon's personal revulsion against poisoning might have influenced his decision. He did not feel that Nixon was deeply committed to environmental preservation.

37. Nixon's ban is in Executive Order 11642, *37 Federal Register*, 2875; the speech is Richard Nixon, "Special Message to the Congress Outlining the 1972 Environmental Program," in *Public Papers of the Presidents of the United States: Richard Nixon, 1972* (Washington, D. C., Government Printing Office, 1974), 173–189. The EPA ban of 9 March is in *37 Federal Register*, 5718. See Cain, "Predator and Pest Control," on legislation and its immediate effects.

Chapter Ten. Saving Species

1. U.S. Congress, House of Representatives, Committee on Merchant Marine and Fisheries, *A Compilation of Federal Laws Relating to Conservation and Development of Our Nation's Fish and Wildlife Resources, Environmental Quality, and Oceanography*

(Washington: Government Printing Office, 1977) (hereafter cited as *Federal Laws*). On legal development, see Michael Bean, *The Evolution of National Wildlife Law* (revised edition; New York: Praeger, 1983).

2. Early history from Endangered Species notebook, Office of Endangered Species, Fish and Wildlife Service. Courtesy of Ronald Nowak (hereafter cited as Endangered Species notebook, OES).

3. Whooping Crane files, Office of Endangered Species, U.S. Fish and Wildlife Service, Washington, D.C. Whooping crane files, WLU 61, Box 5, Records of the Canadian Wildlife Service, Record Group 109, Public Archives Canada, Ottawa (hereafter RG 109, Public Archives Canada, Ottawa).

4. Berryman to Director, Division of Wildlife Research, 8 April 1966, Endangered Species notebook, OES.

5. See, for example, Lewis Regenstein, *The Politics of Extinction: The Shocking Story of the World's Endangered Wildlife* (New York: Macmillan, 1975).

6. Bean, *National Wildlife Law*, 319–321. U.S. Congress, Senate, Committee on Commerce, *Conservation, Protection and Propagation of Endangered Species of Fish and Wildlife*, Hearings before the Merchant Marine and Fisheries Subcommittee of the Committee on Commerce, 89th Congress, 1st Session (Washington: Government Printing Office, 1966). Hereafter all hearings, after their first citation, will be noted by a designation of either House or Senate, a short title, and the year.

7. Steven L. Yaffee, *Prohibitive Policy: Implementing the Federal Endangered Species Act* (Cambridge: MIT Press, 1982), is an excellent source of information on the nuts and bolts of the process of passing and enforcing the endangered species protection program. Yaffee's emphasis on the act as a prohibitive one, though, is questionable; the classification seems arbitrary and overemphasizes one aspect at the expense of others—particularly the programs of research and restoration and the stated need to balance the good from human actions against other values, such as damage to the environment. On the 1968 bill, see pp. 42–45.

8. Senate, *Conservation, Protection (1966); House, Committee on Merchant Marine and Fisheries, Endangered Species*, Hearings before the Subcommittee on Fisheries and Wildlife Conservation of the Committee on Merchant Marine and Fisheries, House of Representatives, 91st Congress, 1st Session (Washington: Government Printing Office, 1969).

9. Bean, *National Wildlife Law*, 321–324; *Endangered Species*, House, 1969.

10. Ronald Nowak, "The Gray Wolf in North America: A Preliminary Report," submitted to the New York Zoological Society and the United States Bureau of Sport Fisheries and Wildlife, 1 March 1974, 221–222. Cited with the author's permission.

11. The territorial legislature first put a bounty on the species in 1915. The following is largely based on Samuel J. Harbo, Jr., and Frederick C. Dean, "Historical and Current Perspectives on Wolf Management in Alaska," in Ludwig N. Carbyn (editor), *Wolves in Canada and Alaska* (Ottawa: Canadian Wildlife Service, 1983), 51–64.

12. Wolf files, 1966, Office of Endangered Species, U.S. Fish and Wildlife Service, Washington, D.C. (hereafter cited as Wolf files, OES). The Fish and Wildlife Service's tentative list of species for restoration is in *Endangered Species*, Senate, 1966, 27–29.

13. "Wolf, Eastern Timber, Corr.," 1965–1966, Wolf files, OES.

14. Wolf files, OES.

15. S. E. Jorgensen to Richard D. Wettersten, 3 December 1969, in "Wolf, East-

ern Timber, Corr.," Wolf files, OES. On Canadian reaction, see correspondence in Wolves, WLU 200, Box 77, RG 109, Public Archives Canada, Ottowa.

16. L. David Mech, *The Wolves of Isle Royale*, Fauna of the National Parks of the United States, Number Seven (Washington: Government Printing Office, 1966). A full account of the Isle Royale work is in Durward L. Allen, *The Wolves of Minong: Their Vital Role in a Wild Community* (Boston: Houghton Mifflin, 1979).

17. Pimott to D. A. Monro, 21 January 1965, in Wolves, WLU 200, RG 109, Public Archives Canada, Ottawa.

18. On radio tracking, see L. David Mech, *Handbook of Animal Radio-Tracking* (Minneapolis: University of Minnesota Press, 1983), and Charles J. Amlaner, Jr., and David W. MacDonald, *A Handbook on Biotelemetry and Radio Tracking* (Oxford: Pergamon Press, 1980). Radio tracking has become a major research technique in the study of many species and a "hot" topic (author's observations at Fourth International Theriological Congress, Edmonton, Alberta, August 1985).

19. Mech, *Handbook*, and "Making the Most of Radio Tracking—A Summary of Wolf Studies in Northeastern Minnesota," in Amlaner and MacDonald, *Handbook of Biotelemetry and Radio Tracking*, 85–95. L. David Mech, *The Wolf: The Ecology and Behavior of an Endangered Species* (Garden City, N.Y.: Natural History Press, 1970). On research support, see "Wolf, Eastern Timber, Corr.," Wolf files, OES.

20. Some taxonomists classified it as a separate species (*Canis niger* at first, then *Canis rufus*), others regarded it as a subspecies of the wolf (*Canis lupus*). R. F. Ewer, *The Carnivores* (Ithaca, N.Y.: Cornell University Press, 1973), 386–387, discusses the question. The latest word is Ronald Nowak's *North American Quaternary Canis* (Lawrence, Kansas: University of Kansas Natural History Museum, 1979).

21. Nowak, "The Gray Wolf," and *Quaternary Canis*. The following discussion is based on these two sources.

22. Stanley Paul Young and Edward A. Goldman, *The Wolves of North America* 1944; reprinted New York: Dover, 1964).

23. Nowak, *Quaternary Canis*, and personal communication. See also U.S. Fish and Wildlife Service, *Red Wolf Recovery Plan* (Atlanta, Georgia: U.S. Fish and Wildlife Service, 1982), and revision, same title and citation, 1984.

24. Richard Nixon, "Special Message to the Congress Outlining the 1972 Environmental Program," in *Public Papers of the Presidents of the United States: Richard Nixon, 1972* (Washington: Government Printing Office, 1974), 173–189.

25. On the endangered species bill, see U.S. Congress, Senate, Committee on Commerce, *Endangered Species Conservation Act of 1972*, Hearings before the Subcommittee on the Environment of the Committee on Commerce, U.S. Senate, 92nd Congress, 2nd Session (Washington: Government Printing Office, 1972). See also Yaffee, *Prohibitive Policy*, 46–57.

26. 16 U.S.C. sec. 1361, in *Federal Laws*, 494. On earlier legal action, see Bean, *National Wildlife Law*, 255–260. On harvesting, see Briton Cooper Busch, *The War Against the Seals* (Kingston: McGill-Queen's University Press, 1985).

27. Bean, *National Wildlife Law*, 283–317; *Federal Laws*, 494–507.

28. House, *Endangered Species* (1973), 7–74; Bean, *National Wildlife Law*, 324–329.

29. House, *Endangered Species* (1973).

30. House, *Endangered Species* (1973); Senate, Committee on Commerce, *Endangered Species Act of 1973*, Hearings before the Subcommittee on the Environment of the Committee on Commerce, U.S. Senate, 93rd Congress, 1st Session (Washington: Government Printing Office, 1973).

31. 16 U.S.C. 1531, 1536; Bean, *National Wildlife Law*, 329–334.

Chapter Eleven. Finding Equilibrium

1. U.S. Congress, House of Representatives, Committee on Agriculture, *Predator Control*, Hearings before the Committee on Agriculture, House of Representatives, 93rd Congress, 1st Session (Washington: Government Printing Office, 1974), 1 (hereafter cited as House, *Predator Control*).

2. This outlook may seem overdrawn, but this position is quite clear in correspondence over a period of fifty years, in official and private files. See Bureau of Biological Survey and Fish and Wildlife Records on predators and the PARC program. See also testimony of National Woolgrowers' Association in U.S. Congress, House of Representatives, Subcommittee on Fisheries and Wildlife Conservation of the Committee on Merchant Marine and Fisheries, *Predatory Mammals and Endangered Species*, Hearings before the Subcommittee on Fisheries and Wildlife Conservation of the Committee on Merchant Marine and Fisheries, U.S. House of Representatives, 92nd Congress, 2nd Session (Washington: Government Printing Office, 1972), 85.

3. "Report of Conferences, 1928–1941," General Files, Division of Wildlife Services, Records of the United States Fish and Wildlife Service, Record Group 22, National Archives, Washington, D.C. (hereafter cited as RG 22, National Archives).

4. U.S. Congress, House, *Predator Control* (1973), 12–23.

5. Executive Order 11870, 18 July 1975, copy from National Wildlife Federation.

6. Kimball to Gerald Ford, 20 May 1975 and 25 July 1975, correspondence and announcements of the National Wildlife Federation; author's interview with Thomas Kimball, 19 December 1979.

7. See, for example, Ray Coppinger, Jay Lorenz, John Glendining, and Peter Pinardi, "Attentiveness of Guarding Dogs for Reducing Predation on Domestic Sheep," *Journal of Range Management, 36* (May 1983), 275–279. Coppinger spoke on his continuing research at the International Wolf Symposium sponsored by Defenders of Wildlife, 22 May 1987, Washington, D.C. (hereafter Defenders Wolf Symposium, 1987). Author's notes taken at the meeting.

8. The files of various organizations interested in humane legislation and game preservation have much material on this. For summaries of some recent work, see Walter E. Howard, Roy Teranishi, Rex E. Marsh, and Jerry H. Scrivner, "Understanding Coyote Behavior," *California Agriculture* (March-April 1985), 4–7; Daniel B. Fagre and Roy Teranishi, "Development of Coyote Attractants for Reduction of Livestock Losses," typescript of paper from "Proceedings of the Wildlife-Livestock Relationships Symposium" (Moscow, Idaho: Forest, Wildlife & Range Experiment Station, 1982), 319–326.

9. U.S. Department of the Interior, Fish and Wildlife Service, *Predator Damage in the West: A Study of Coyote Management Alternatives* (Washington: U.S. Fish and Wildlife Service, 1978), 115.

10. U.S. Department of the Interior, Office of Audit and Investigation, *Review of the Animal Damage Control Program, U.S. Fish and Wildlife Program*, November 1978. Quote is from p. 6.

11. U.S. Department of the Interior, Fish and Wildlife Service, "Draft Environmental Impact Statement, Animal Damage Control Program" (Washington: U.S. Department of the Interior, 1978).

12. Barry Flamm to Lynn Greenwalt, 12 February 1979. Copy from files of National Wildlife Federation.

13. Roman H. Koenings to Lynn Greenwalt, 27 March 1979, Bureau of Land

Management analysis of draft Environmental Impact Statement. Copies from National Wildlife Federation.

14. Jane Yarn to Robert L. Herbst, 12 February 1979. Copy from National Wildlife Federation.

15. John Grandy to Cecil Andrus, 15 March 1979. Copy from National Wildlife Federation.

16. Stahr and Kimball to Andrus, 14 February 1979. Copy from National Wildlife Federation.

17. U.S. Fish and Wildlife Service, Department of the Interior, "Final Environmental Impact Statement, Animal Damage Control Program" (Washington: U.S. Department of the Interior, 1979).

18. Advisory Board on Wildlife Management, "Predator and Rodent Control in the United States," in *Transactions of the Twenty-ninth North American Wildlife and Natural Resources Conference* (Washington: Government Printing Office, 1964), 6.

19. Cecil D. Andrus, "Remarks of the Honorable Cecil D. Andrus, Secretary of the Interior, to the Conference on Animal Damage Control," Austin, Texas, 15 January 1980. Copy from the National Wildlife Federation.

20. James Watt to Robert Jantzen, Acting Director, Fish and Wildlife, 22 September 1981.

21. *40 Federal Register*, 7 December 1981, 59622–59630.

22. Supporting organizations were the National Audubon Society, the Humane Society of the United States, the American Humane Association, the Animal Protective Institute of America, the National Parks and Conservation Association, the Animal Welfare Institute, the Fund for Animals, Natural Resources Defense Council, the Sierra Club, Friends of the Earth, and the Environmental Defense Fund. The National Wildlife Federation, though it did not join the coalition, generally agreed with it.

23. U.S. Environmental Protection Agency, "In the Matter of Notice of Hearing on the Applications to Use Sodium Fluoroacetate (Compound 1080) to control Predators," FIFRA Docket No. 502, Final Decision, 31 October 1983, by Lee M. Thomas. Copy from Defenders of Wildlife.

24. Stephen R. Kellert, *Public Attitudes toward Critical Wildlife and Natural Habitat Issues* (Washington: Government Printing Office, 1979), 52–75, discusses public attitudes and needs. This study has all the usual flaws of public-opinion surveys, but it is roughly in accord with what the political record shows.

25. L. David Mech and Robert Rausch, "The Status of the Wolf in the United States, 1973," in Douglas H. Pimlott (editor), *Wolves: Proceedings of the First Working Meeting of Wolf Specialists and of the First International Conference on the Conservation of the Wolf* (Morges, Switzerland: International Union for the Conservation of Nature and Natural Resources, 1975), 78.

26. The following is largely based on Samuel J. Harbo, Jr., and Frederick C. Dean, "Historical and Current Perspectives on Wolf Management in Alaska," in Ludwig N. Carbyn (editor), *Wolves in Canada and Alaska* (Ottawa: Canadian Wildlife Service, 1983), 51–64. Quote from p. 51.

27. Harbo and Dean, "Wolf Management in Alaska," 63.

28. Eastern Timber Wolf Recovery Team, *Recovery Plan for the Eastern Timber Wolf* (Washington: Government Printing Office, 1978), Preface.

29. Eastern Timber Wolf, *Recovery Plan*, 4–6.

30. Eastern Timber Wolf, *Recovery Plan*, Appendix D.

31. All letters are in Eastern Timber Wolf, *Recovery Plan*, Volume Two. Those

joining Defenders were Animal Welfare Institute, Aubudon Naturalist Society of the Central Atlantic States, Inc., Friends of the Earth, International Fund for Animal Welfare, National Parks and Conservation Association, Sierra Club, The Fund for Animals, The Wilderness Society, and Wild Canid Survival and Research Center.

32. Jack Ohman to Ralph Bailey, Eastern Timber Wolf Recovery Team leader, 26 January 1976, in Eastern Timber Wolf, *Recovery Plan*, Volume Two (unpaged).

33. Wolf file, General Files, Office of Endangered Species, U.S. Fish and Wildlife Service, Washington, D.C.

34. Wolf file, OES. Notice of change of classification from "endangered" to "threatened" in *43 Federal Register*, 8 March 1978, 9607–9615.

35. U.S. Fish and Wildlife Service, *Northern Rocky Mountain Wolf Recovery Plan* (Denver: U.S. Fish and Wildlife Service, 1980).

36. U.S. Fish and Wildlife Service, *Northern Rocky Mountain Wolf Recovery Plan*, 30–31.

37. U.S. Fish and Wildlife Service, *Red Wolf Recovery Plan* (Atlanta: U.S. Fish and Wildlife Service, 1982).

Epilogue

1. Douglas H. Pimlott, "Wolf Control in Canada," *Canadian Audubon* (November-December 1961), 2–9. Reprinted by the Canadian Wildlife Service, Department of Indian Affairs and Northern Development. Copy courtesy Ronald Nowak, Office of Endangered Species, Washington, D.C.

2. John B. Theberge, "Wolf Management in Canada Through a Decade of Change," *Nature Canada*, 2 (January 1973). Unpaginated reprint by Canadian Nature Federation. Copy courtesy Ronald Nowak, Office of Endangered Species, Washington, D.C.

3. Jim Robbins, "Wolves across the Border," *Natural History*, 95 (May 1986), 6–15; Jack Horan, "The Red Wolf is Coming Home," *Defenders* (May-June 1986), 4–11; Warren T. Parker, "A Technical Proposal to Reestablish the Red Wolf on Alligator River National Wildlife Refuge, N.C." (Asheville, North Carolina: U.S. Fish and Wildlife Service, 1986); Warren T. Parker, Project Leader, Red Wolf Team, personal communication, 30 December 1986; U.S. Fish and Wildlife Service, *Northern Rocky Mountain Wolf Recovery Plan* (Denver: U.S. Fish and Wildlife Service, 1980); idem, *Agency Review Draft, Revised Northern Rocky Mountain Wolf Recovery Plan* (Denver: U.S. Fish and Wildlife Service, 1984); idem, *Red Wolf Recovery Plan* (Atlanta: U.S. Fish and Wildlife Service, 1982); idem, *Red Wolf Recovery Plan* (Atlanta: U.S. Fish and Wildlife Service, 1984); Robert Ream, Northern Rocky Mountain Wolf Recovery Team, "The Migration of Wolves to Montana: Solving Problems Facing Natural Recovery," and Warren Parker, Red Wolf Recovery Team, "Reintroduction of the Red Wolf—Formula for Success," International Wolf Symposium sponsored by Defenders of Wildlife, 22 May 1987, Washington, D.C., author's notes (hereafter Defenders Wolf Symposium, 1987).

4. Manifesto in Douglas H. Pimlott (editor), *Wolves: Proceedings of the First Working Meeting of Wolf Specialists* (Morges, Switzerland: International Union for Conservation of Nature and Natural Resources, 1975).

5. "Notes on the Working Meeting: Conservation of the Wolf in Europe," in Pimlott (editor), *Wolves*, 9–11; Fred H. Harrington and Paul C. Paquet (editors), *Wolves of the World* (Park Ridge, N.J.: Noyes Publications, 1982).

6. Luigi Boitani, "The Wolf: A Global Perspective," Defenders Wolf Symposium, 1987.

7. The eastern coyote has some wolf genes, probably from matings with wolves in disturbed habitat in eastern Canada, but the animals are still quite distinctly coyotes. See Ronald Nowak, *Quaternary Canis* (Lawrence, Kansas: Museum of Natural History, 1979), 21. Local Virginia news and newspapers, for example the *Roanoke Times and World-News*, have carried reports of coyotes and coyotelike animals in the last few years.

8. Discussion at meeting of the Canid Specialist Group, Fourth International Theriological Congress, Edmonton, Alberta, August 1985, author's notes.

9. *Wall Street Journal*, 20 October 1981. Canid Specialist Group meeting, Edmonton, author's notes.

10. Clarence Glacken, *Traces on the Rhodian Shore* (Berkeley: University of California Press, 1976); Keith Thomas, *Man and the Natural World* (New York: Pantheon, 1983); Roderick Nash, *Wilderness and the American Mind*, 3rd edition (New Haven: Yale University Press, 1983).

11. Defenders Wolf Symposium, 1987.

Index

Library of Congress Cataloging-in-Publication Data

Dunlap, Thomas R., 1943–
Saving America's wildlife.

Includes index.
1. Wildlife conservation—United States—History. I. Title.

QL84.2.D86 1988   333.95′16′0973   87–25940
ISBN 0–691–04750–2 (alk. paper)

NEW ROCHELLE PUBLIC LIBRARY

3 1019 15001638 7

333.95 D
Dunlap, Thomas R.,
Saving America's wildlife
$24.95

JUN 10 1993

MAR 15 1994

MAR 25 1995

APR 17 1995

1/89

NEW ROCHELLE PUBLIC LIBRARY

Library Plaza

New Rochelle, N.Y.   10801

632-7878

Please telephone for information

on library services and hours